ESTUARINE ECOHYDROLOGY

This book is a result of, and a contribution to, the UNESCO ecohydrology programme, which is a component of IHP-VI. It has been made possible as a result of my collaboration with the European Regional Centre for Ecohydrology, under the auspices of UNESCO, Lódz, Poland. This book is also endorsed by Japan-based EMECS (International Center for Environmental Management of Enclosed Coastal Seas) and SCOR (Scientific Committee on Oceanic Research).

Cover photo: Infrared aerial photo of the mouth of the Parker River and Northern Plum Island Sound, Newbury, Massachusetts, U.S.A, 1:25,000 scale.

ESTUARINE ECOHYDROLOGY

ERIC WOLANSKI, PhD, DSc, FTSE, FIE Aust
Australian Institute of Marine Science
Townsville
Australia

ELSEVIER

Amsterdam • Boston • Heidelberg • London • New York • Oxford
Paris • San Diego • San Francisco • Singapore • Sydney • Tokyo

Elsevier
Radarweg 29, PO Box 211, 1000 AE Amsterdam, The Netherlands
Linacre House, Jordan Hill, Oxford OX2 8DP, UK

First edition 2007

Notice
No responsibility is assumed by the publisher for any injury and/or damage to persons
or property as a matter of products liability, negligence or otherwise, or from any use
or operation of any methods, products, instructions or ideas contained in the material
herein. Because of rapid advances in the medical sciences, in particular, independent
verification of diagnoses and drug dosages should be made

Library of Congress Cataloging-in-Publication Data
A catalog record for this book is available from the Library of Congress

British Library Cataloguing in Publication Data
A catalogue record for this book is available from the British Library

ISBN: 978-0-444-53066-0

For information on all Elsevier publications
visit our website at books.elsevier.com

Printed and bound in The Netherlands

07 08 09 10 11   10 9 8 7 6 5 4 3 2 1

ABOUT THE AUTHOR

Dr. Eric Wolanski, PhD, DSc, FTSE, FIE Aust, is a coastal oceanographer and a leading scientist at the Australian Institute of Marine Science. He obtained a B.Sc. degree in civil engineering from the Catholic University of Louvain, a M.Sc. degree in civil and geological engineering from Princeton University, and a Ph.D. in environmental engineering from The Johns Hopkins University.

His research interests range from the oceanography of coral reefs, mangroves, and muddy estuaries, to the interaction between physical and biological processes determining ecosystem health in tropical waters. He has more than 300 publications.

He is a fellow of the Australian Academy of Technological Sciences and Engineering, the Institution of Engineers Australia, and l'Académie Royale des Sciences d'Outre-Mer. He was awarded an Australian Centenary medal for services in estuarine and coastal oceanography, a Doctorate Honoris Causa from the Catholic University of Louvain, and a Queensland Information Technology and Telecommunication award for excellence.

He is the chief editor of *Estuarine, Coastal and Shelf Science* and *Wetlands Ecology and Management*. He is a member of the editorial board of *Journal of Coastal Research*, *Journal of Marine Systems*, and *Continental Shelf Research*. He is a member of the Scientific and Policy Committee of the Japan's-based International Center for Environmental Management of Enclosed Coastal Seas. He is an Erasmus Mundus scholar. He is listed in Australia's Who's Who.

CONTENTS

CHAPTER 1

Introduction

1.1. WHAT IS AN ESTUARY?

An estuary receives, occasionally or frequently, an inflow of both freshwater and saltwater; it stores these waters temporarily while mixing them. An estuary is a buffer zone between river (freshwater) and ocean (saltwater) environments that may be affected by tidal oscillations. Most estuaries were established by the flooding of river-eroded or glacially-scoured valleys during the Holocene rise of sea level starting about 10,000–12,000 years ago. Because it progressively fills with sediment, an estuary has an age, akin to a living organism in evolution. It starts with youth, it matures, and it then becomes old; it can be rejuvenated.

The term "estuary" is derived from the Latin word "aestuarium", this means tidal. This definition is however over-restrictive because estuaries also occur in conditions with no tides such as the rivers discharging into the tideless Baltic Sea and the Danube Delta in the tideless Black Sea.

There have been several definitions of an estuary, such as that of Dionne (1963): "*An estuary is an inlet of the sea, reaching into the river valley as far as the upper limit of tidal rise, usually divisible into three sectors: a) a marine or lower estuary, in free connection with the open sea; d) a middle estuary, subject to strong salt and freshwater mixing; and c) an upper or fluvial estuary, characterized by freshwater but subject to daily tidal activity*".

There are other definitions of an estuary. For instance Pritchard (1967) defined an estuary as "*a semi-enclosed coastal body of water, which has a free connection with the open sea, and within which sea water is measurably diluted with fresh water derived from land drainage*". This definition based on salinity has been accepted for the last forty years. It excludes a number of coastal water bodies such as hyper-saline tropical lagoons with no perennial inflows; it also seems to exclude seasonally closed lagoons; it explicitly excludes the Baltic Sea, the Seto Inland Sea in Japan, and other brackish seas.

Dalrymple et al. (1992) proposed a definition of an estuary from the point of view of the fluvial and marine sources of the sediment. Perillo (1995) offered another definition based on the dilution of freshwater with seawater and the presence of euryhaline biological species. An estuary can also be defined as the zone stretching from the tidal limit to the seaward edge of the tidal plume in the open ocean (Kjerfve, 1989).

1

The definition of an estuary in this book combines all these definitions, i.e. an estuary is a semi-enclosed body of water connected to the sea as far as the tidal limit or the salt intrusion limit and receiving freshwater runoff, recognizing that the freshwater inflow may not be perennial (i.e. it may occur only for part of the year) and that the connection to the sea may be closed for part of the year (e.g. by a sand bar) and that the tidal influence may be negligible. The definition includes fjords, fjards, river mouths, deltas, rias, lagoons, tidal creeks, as well as the more classical estuaries. It recognises commonalities with predominantly brackish areas such as the Baltic Sea, and freshwater-poor coastal waters in arid zones.

It is also difficult to define where an estuary ends. This is usually assumed to be an abrupt coastline break. However many estuaries change shape gradually, thus the transition between river, estuary, coastal embayment and open coast is gradual and not always obvious. To accommodate these problems, the European Environmental Agency coined the word 'Transitional Waters' as *"bodies of surface water in the vicinity of river mouths which are partly saline in character as a result of their proximity to coastal waters but which are substantially influenced by freshwater flows"*. Some of these 'transitional waters' are not river mouths nor have substantially lowered salinity, and they are neither rivers nor open coasts. These are considered as 'estuaries' in this book.

1.2. HUMANITY AND ESTUARIES

Estuaries and continental shelf areas comprise 5.2% of the earth surface, and only 2% of the oceans' volume. However, they carry a disproportionate human load. At present, about 60% of the world's population lives along the estuaries and the coast (Lindeboom, 2002). Throughout human history, estuaries have been amongst the most populated areas over the world. This is because people used them as transport routes, and because of their high biological productivity sustaining a high level of food production; indeed, coastal waters supply about 90% of the global fish catch (Wolanski et al., 2004a).

The human population worldwide is presently doubling every 30–50 years (Fig. 1.1a) and this increase is unprecedented in human history (Fig. 1.1b). Because of internal migration of people away from the hinterland towards the coast, the population is now doubling every 20 years along many coasts.

The toll for estuaries and coastal waters is severe; more and more they are unable to sustain the quality of life that people searched for when migrating to the coast in the first place. The reasons for this degradation are many. They include,

1.2.1. Sedimentation from sediment eroded from cleared land in the hinterland

Deforestation, overgrazing and other poor farming practices, as well as roads and mining, increase soil erosion (Fig. 1.2 a–d) and the sediment loads in rivers

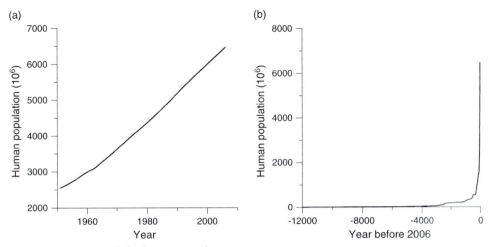

FIGURE 1.1. Growth of the human population.
Source: U.S. Bureau of the Census, International Data Base.

FIGURE 1.2. Photographs of (a) land clearing to the edge of a small estuary draining into the Great Barrier Reef of Australia, (b) cattle overgrazing the banks of the Ord River estuary, Australia, (c) massive land slide from the discharge of mining waste in the upper Fly River estuary, Papua New Guinea, (d) massive mud flow smothering the coastal plains from a land slide in the Philippines, likely initiated by deforestation over steep slopes. Photos (a) and (d) are courtesy of V. Veitch and J. Ramirez.

TABLE 1.1. Comparison of the drainage areas, the sediment load and the yield for various rivers.

River	Area (10^6 km^2)	Yield (tonne km^{-2} year^{-1})
Minimal land use		
Ngerdoch (Palau)	39×10^{-6}	2
King Sound (Australia)	0.12	50
Moderate land use		
Yangtze (China)	1.9	252
Amazon (Brazil)	6.1	190
Mississippi (U.S.A.)	3.3	120
Mekong	0.79	215
Extensive land use		
La Sa Fua (Guam)	5×10^{-6}	480
Ganges/Brahmaputra (India)	1.48	1670
Cimanuk (Indonesia)	0.0036	6350

The effect of deforestation on estuaries is much more rapid in the tropics than in temperate zones because of intense rainfall in the tropics. The catchments of the Cimanuk and La Sa Fua rivers are small and profoundly modified by human activities. The Ngerdoch River drains a hilly, forested area. The sediment yield is largely determined by the climate, the topography and human activities, and is weakly dependent on the catchment size. (Data from Wolanski and Spagnol, 2000; Syvitski et al., 2005; Victor et al., 2005).

by typically a factor of 10 largely independently of catchment size and mainly dependent on the degree of land clearing in the mountainous part of the river catchment (Table 1.1).

As a result, the shoreline can change from sandy to muddy, diminishing the quality of life of the population on its shore (Fig. 1.3). Increased muddiness and turbidity in the estuary result. This smothers the benthos (Fig. 1.4a) and degrades the ecosystem by decreasing the light available for photosynthesis. This degradation is further increased by dredging (Fig. 1.4b) and dumping of dredged mud within the estuary (Fig. 1.4c).

Land clearing also increases peak flood flows by up to 30% and decreases dry season flows, thus exacerbating flooding in the wet season and droughts in the dry season (Wolanski and Spagnol, 2000).

1.2.2. Overfishing and trawling

This muddies the water, destroys fish stock and damages or destroys the benthos and habitats (Fig. 1.4d; Trimmer et al., 2005; Jennings et al., 2001; Rijnsdorp et al., 1998). The decline of coastal fish stocks has been dramatic worldwide. As an example, the demersal fisheries biomass in Manila Bay has decreased from 8,290 tons in 1947 to 840 tons in 1993 (Jacinto et al., 2006). As another example, the declared catch of the Beluga sturgeon in the Danube River, a fish migrating upriver from offshore waters, has decreased from 400–600 tons in 1950, 200 tons in 1974, and 10 tons in 2005 (I. Jelev, pers. com.)

FIGURE 1.3. The coastline of Cairns, Australia, has changed in one century from sandy to muddy as a result of land clearing in the rivers' catchment. Adapted from Wolanski and Duke (2002).

1.2.3. Destruction of wetlands

Wetlands are infilled for harbours and marinas (Fig. 1.5a), urbanisation including slums (Fig. 1.5b), garbage disposal (Fig. 1.5c), aquaculture (Fig. 1.5d), and dykes for farming (Fig. 1.5e). Nearly all estuarine marshes have already been reclaimed in Japan and in The Netherlands, with resulting loss of organic material to oxidation and burning as well as a decrease of soil elevation, a major cause of flooding. Drainage of wetlands can lead to acidification problems that result in vegetation and fish kills (Fig. 1.5f; Soukup and Portnoy, 1986).

1.2.4. Eutrophication

This is the water quality degradation caused by excessive nutrients; much of it is derived from sewage (Fig. 1.6a) and animal waste from agricultural feedlots (Fig. 1.6b). Eutrophicated waters suffer from a significant reduction in dissolved oxygen leading to hypoxia (dissolved oxygen concentration, DO, $< 2\,\mathrm{mg\,l^{-1}}$) and anoxia (DO $= 0$; Diaz and Rosenberg, 1995; Richardson and Jorgensen, 1996).

FIGURE 1.4. Photographs of (a) a coral reef smothered and killed by eroded soil as a result of land clearing, Airai Bay, Palau, (b) the sediment plume in the lee of a dredger, Townsville, Australia, (c) discharging dredged mud, Singapore, (d) habitat destruction and the sediment plume from a trawl net, Great Barrier Reef of Australia. Photo (a) was provided by R.H. Richmond. Photo (d) is modified from Wolanski (2001).

The benthos and resident fauna die when the DO is less than $1\,mg\,l^{-1}$. When hypoxia occurs, fish, crabs, and shrimp attempt to migrate away. Human-induced hypoxia is widespread in coastal waters worldwide. In the Gulf of Mexico, the discharge of the Mississippi River creates a 'dead zone', i.e. a bottom-tagging, hypoxic water zone ($DO \leq 2\,mg\,l^{-1}$) of 8,000–9,000 km^2 in 1985 to 1992, and increasing to 16,000–20,000 km^2 in 1993-2000 (Rabalais et al., 2002). Hypoxic waters cover 84,000 km^2 of the Baltic Sea and 40,000 km^2 of the northwestern shelf of the Black Sea where historically hypoxia existed but anoxic events became more frequent and widespread in the 1970s and 1980s. Declines in bottom-water dissolved oxygen have been reported for the northern Adriatic Sea, the Kattegat and Skaggerak, Chesapeake Bay, the German Bight and the North Sea, Long Island Sound in New York, as well as numerous other estuaries worldwide. Additionally nuisance or harmful algae blooms (HABs) are now common in many estuaries and coastal seas (Fig. 1.6c; Graneli and Turner, 2006). At times this results in severe deficits of dissolved oxygen, leading to hypoxia and anoxia and destruction of whole ecosystems as exemplified by massive fish kills (Fig. 1.6d).

The human activities degrading the Black Sea ecosystem and the Gulf of Mexico are not just located on the coast. Instead they occur throughout the drainage basin of, respectively, the Danube River that drains eight European countries, and the Mississippi River that drains much of the United States (Rabalais et al., 2002; Richardson and Jorgensen, 1996; Zaitsev, 1992; Lancelot

FIGURE 1.5. Photographs of (a) urbanization that destroyed wetlands of the Coomera River estuary, Gold Coast, Australia, (b) slums that have infilled wetlands along the Mekong River estuary, Vietnam, (c) a sarcocornia salt marsh near petrochemical industry in Bahia Blanca, Argentina, that is used for garbage dumping, (d) the wet desert creating by dead and dying shrimp ponds that have destroyed mangrove forests in Chumpon, Thailand, (e) a dyke that destroyed 70% of a mangrove swamp in Trinity Island, Cairns, Australia, (f) a fish kill from acid leachate from drainage of a wetland near Ingham, Australia. Photos (a), (c) and (f) are courtesy of N. Duke, R. Lara and V. Veitch.

et al., 2002). A similar degradation from human land-based activities is also observed in poorly flushed embayments of Japan, including the Seto Inland Sea, Osaka Bay, and Tokyo Bay, where HABs occur 100 days per year (Okaichi and Yanagi, 1997; Takahashi et al., 2000; Furukawa and Okada, 2006).

1.2.7. Dykes for flood protection

These shut off the natural flood plains from the rivers and estuaries. The estuarine ecosystem is degraded by increased floods, both height and discharge, as a result of the removal of the buffering effect of the flood plains.

1.2.8. Human health risks

The degradation of estuaries leads to threats to human health (Table 1.2).

TABLE 1.2. Examples of threats to human health arising from the degradation of estuarine ecosystems.

Driving force	Changing ecological pattern	Influence on human health
Pollution from oil, industry, naval operations and sewage discharge	Deterioration of marine ecosystems from imbalances due to dense ship traffic	Decrease in life expectancy, skin and eye diseases (Black/Azov seas, Caspian Sea)
Biological/bacterial contamination due tohydrological changes	Effects on fish and algae	Typhoid, malaria, diphtheria (Central Asia)
Biological contamination of surface water with waste water		Gastroenteritis, eye and skin infections (UK, France, S. Africa)
Biological and chemical contamination: harmful (toxic and nontoxic) algal blooms from the rapid reproduction and localized dominance of phytoplankton	Shellfish poisoning, wildlife mortalities, sunlight penetration prevention, oxygen shortages, reservoirs for bacteria	Poisonings, diarrhea, dehydration, headaches, confusion, dizziness, memory loss, weakness, gastroenteritis, bacterial infections, swimming-related illnesses, neurological diseases, deaths (Florida, Gulf States, S. America)
Contamination: cholera-contaminated sea plankton due to contaminated ships' hulls	Coastal shellfish and fish contamination	Cholera (Peru and other 16 countries)

Modified from UNEP (1999).

1.3. THE FUTURE OF ESTUARIES AND THE QUALITY OF LIFE OF THE HUMAN POPULATION LIVING ON ITS SHORES

The people living near estuaries and the coast are experiencing increasing degradation of estuarine water quality and decreasing ecological services delivered by the estuary. Simultaneously their expectations of a high quality of life are increasingly unrealistic and unfulfilled in view of reduced biodiversity, reduced productivity, and reduced health of estuaries and coastal waters. Humans are increasingly moving away from the possibility of ecologically-sustainable development of the coastal zone.

The human impact on the ecological health of estuaries depends on several factors, one of them is water circulation. The water circulation in some estuaries is swift and readily flushes away pollutants to the open ocean; other estuaries are poorly flushed and the pollutants are retained. Unfortunately many estuarine environments that are especially attractive for human settlements, such as wetlands, lagoons, harbours, and fjords, are often poorly flushed and are thus more prone to pollution or degradation. If the extra load of nutrients and pollutants is small enough and the estuary is rapidly flushed, the biological productivity is boosted without dramatically modifying the biodiversity (Zalewski, 2002). When the extra load of nutrients and pollutants is high or the estuary is slowly flushed, the estuary is degraded in terms of water quality, ecological services, and biodiversity. This degradation is not restricted to estuaries, it can extend to coastal waters if these are also poorly flushed. Eutrophication and in the worst cases anoxia and/or toxic algae blooms (red tides) can result, e.g. the Baltic Sea in Europe and the Pearl Estuary in China (Gren et al., 2000).

Better flushed and larger systems also suffer from environmental degradation through eutrophication, as is made apparent by some beaches of the North Sea being covered by foams of decaying algae and protozoa, mainly *Phaeocystis* and *Noctiluca* (Lindeboom, 2002).

In extreme cases, estuaries have essentially become industrial drains that are also used for navigation. Such is the case of the Pearl Estuary in China and the Saigon River in Vietnam (Wolanski, 2006a).

This degradation extends in coastal waters to seagrass and coral reefs. The percentage of dying reefs is highest in countries with widespread land clearing (50% in Taiwan and Vietnam, 35% in the Philippines; Bourke et al., 2002). On a global scale, reefs are as much threatened by pollution from land runoff as by coral bleaching due to global warming, overfishing and destructive fishing (Fabricius, 2005; Spalding et al., 2001).

Until now, the solution was believed to depend on reducing the amount of waste and relying on hard technology, namely the construction of sewage treatment plants and the modification of farming practices and technology. While this has restored some ecological functions in the Rhine and Thames estuaries, this technological fix has not restored the ecological health of estuaries in both developed and developing nations worldwide.

The reasons for failure are simple. *Firstly*, integrated coastal zone management plans are drawn up but, in the presence of significant river input, they

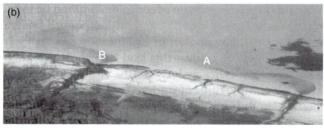

FIGURE 2.5. (a) Oblique aerial photograph of the wetlands of Missionary Bay, Australia, showing the extensive network of mangrove-fringed tidal creeks. On exiting one creek at falling tide, water flows out to the bay, and re-enters that creek or another creek at the following rising tide. This process leads to a long (50 days) exposure time for the bay while the residence time in a creek is 7–10 days. (b) This aerial photograph of the coast of the Gulf of Carpentaria, Australia, shows that the water 'A' leaving the creek on the right is deflected by the longshore current and can be re-entrained into creek B at the next rising tide. This increases the value of the return coefficient of the system.

Except in microtidal estuaries or during river floods for meso- and micro-tidal estuaries, the total fluxes through the river mouth of brackish water during the rising and falling tides are commonly much higher, often by a factor of 10 to 100, than the volume flux due to the riverine inflow. The net flux is the small difference between two large numbers. If the measurements are not extremely precise, the resulting estimate of the net flux is unreliable. Hence the direct measurement of the return coefficient is rarely successful. In practice it is impossible to directly measure the return coefficient in systems flushed by unsteady events in coastal waters such as an upwelling, the passage of an oceanic eddy, and storms because these events occur at time scales much longer than the tidal time scale during which field measurements are based (de Castro et al., 2006; Hinata, 2006).

2.4. VERTICAL MIXING AND STRATIFICATION

The residence time of water in an estuary also depends on the internal circulation within the estuary driven by the density difference generated by changes in salinity and temperature. Freshwater floats over saline water, and warmer water floats over colder water for temperatures greater than about 4 °C. Thus near-bottom and near-surface waters can have different trajectories, and thus different residence times.

Vertical mixing (Fig. 2.6a) determines how much the salinity and the temperature change from top to bottom. This profoundly affects the water circulation. Vertical mixing occurs from the surface downward forced by the wind, from the bottom upward forced by boundary-generated turbulence (boundary mixing), and internally by turbulent mixing driven by the water currents driven by tides, wind and river inflow. Vertical mixing is parameterized by the vertical eddy diffusion coefficient K_z. In vertically well-mixed conditions, i.e. where differences in salinity or temperature between the top and the bottom are negligible, K_z is maximum in mid-waters. In the presence of a strong stratification in temperature, salinity, or suspended sediment concentration, a density interface exists. Its general name is pycnocline. It is also called a salinity step structure, a thermocline, and a lutocline if the density difference is due to, respectively, salinity, temperature, or suspended sediment concentration. K_z is the smallest at the level of the density interface (i.e. the pycnocline; Fig. 2.6b), due to buoyancy effects inhibiting mixing.

Additional mixing also results from turbulence generated at the left and right banks of the estuary (Fischer et al., 1979). Additional mixing is provided by secondary flows around shoals and islands, as well as flows in meanders.

Mixing generates different types of estuarine circulation. A partially stratified estuary (Fig. 2.7a) has isohalines that are sloping smoothly from top to bottom.

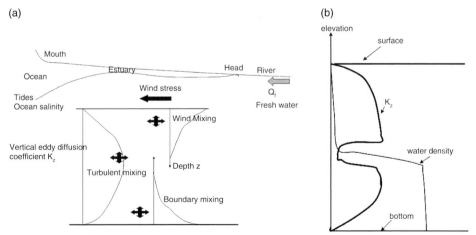

FIGURE 2.6. (a) Sketch of the vertical mixing processes in a vertically fairly well-mixed estuary. K_z is the vertical eddy diffusion coefficient. (b) Vertical profile of K_z in a stratified estuary where the water density gradient can be due to temperature, salinity, and suspended sediment concentration.

(a)

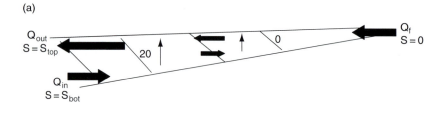

(b)

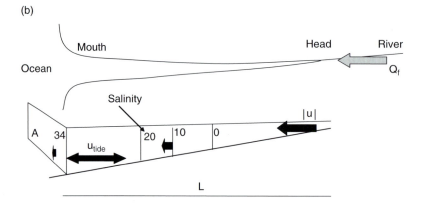

(c)

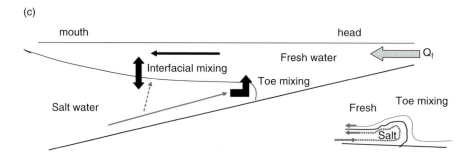

(d)

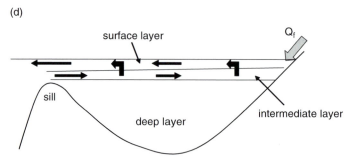

FIGURE 2.7. Sketch of the water circulation in (a) a partially stratified estuary, (b) a vertically well-mixed estuary, (c) a salt-wedge estuary including details of the flow in the toe of the salt wedge, (d) a fjord. In (a) and (b), the numbers are salinity values.

The water circulation is characterized at the mouth by a near-bottom inflow of oceanic water and a near-surface outflow of brackish water. The near-bottom inflow prevails all the way from the mouth to the salinity intrusion limit. The internal circulation Q_{in} is calculated using Eqs. (2.1) and (2.2).

A vertically well-mixed estuary (Fig. 2.7b) has vertical isohalines. The salinity introduces negligible currents. The mean velocity $|u|$

$$|u| = Q_f/A \qquad (2.11)$$

where Q_f is the riverine discharge and A the cross-section of the estuary, leads to the estimate of a mean advective residence time T_a,

$$T_a = L/|u| \qquad (2.12)$$

T_a can be a reliable estimate of the residence time only for micro-tidal estuaries and lagoons. T_a greatly overestimates the residence time in macro-tidal estuaries with swift tidal currents.

A salt-wedge estuary (Fig. 2.7c) is characterized by a very sharp density interface between freshwater on top and saline water on the bottom. There is usually intense mixing at the toe of the salt wedge, where the freshwater plume lifts off the bottom, and this results in entrainment of saline water into the plume. The shear between the two water masses generates a return flow within the salt wedge. There is also interfacial mixing further downstream of the toe; this is usually weak but, spread over long distances, it results ultimately in the mixing between riverine and ocean waters.

Fjords (Fig. 2.7d) are silled basins with freshwater inflow greatly exceeding evaporation. There is an import of oceanic water in an intermediate layer. This water mixes with the freshwater inflow and the brackish water from that mixing is exported in the surface layer. There may be a slow import of saline water over the sill, this water sinks into the bottom of the fjord, i.e. the deep layer, where water stagnates until occasionally flushed by storms.

Inverse estuaries (Fig. 2.8a) occur in dry climates including the tropics in the dry season, when evaporation greatly exceeds freshwater inflow. A salinity maximum zone is formed and both oceanic and riverine water flow near the surface towards this zone (Wolanski, 1986). There, this water downwells and spreads along the bottom both seaward and landward. The rate of water loss by evaporation is,

$$E_i = e\,W_i \qquad (2.13)$$

where e is the free water evaporation rate and W_i is the surface area of the water body ($i = 1$ in the estuary; $i = 2$ in the salinity maximum zone). The equations for conservation of water and salt in the estuary upstream from the salinity maximum zone are,

$$Q_f + Q_2 = Q_1 + E_1 \qquad (2.14)$$

$$Q_1 S_2 = Q_2 S_3 \qquad (2.15)$$

If $Q_f > E_1$, a classical estuarine circulation prevails upstream of the salinity maximum zone. Conservation of water and salt for the estuary including the salinity maximum zone requires,

$$Q_f + Q_3 = Q_4 + E_1 + E_2 \qquad (2.16)$$

$$Q_3S_1 = Q_4S_0 \tag{2.17}$$

From Eqs. (2.15) and (2.16),

$$Q_3 = (E_1 + E_2 - Q_f)/(1 - S_1/S_0) \tag{2.18}$$

If $E_1 + E_2 > Q_f$, an inverse estuarine circulation prevails downstream of the salinity maximum zone. Thus water escapes from the salinity maximum zone by downwelling and flows away along the bottom.

The maximum salinity reaches extremely high values, e.g. 100 in estuaries in Senegal and 40 in Australia (Wolanski, 1986; Pages et al., 1995). The residence time is several months. In those systems, the salinity maximum zone is a plug that inhibits the mixing of estuarine and ocean waters so much that freshwater does not reach the ocean. High salinity water leaves the estuary and sinks seaward (Nunes and Lennon, 1986; de Castro et al., 2004; Ribbe, 2006).

Tidal creeks in the tropics in the dry season behave as evaporation ponds (Fig. 2.8b). These creeks are usually fringed by mangroves, salt marshes, or salt pans. Evaporation occurs at a rate e over the whole area W of the creek and the

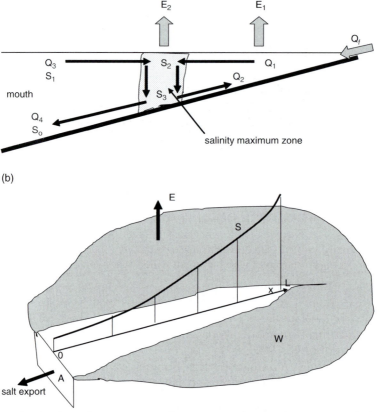

FIGURE 2.8. Sketch of circulation patterns characteristic of tropical estuaries, including (a) the formation of a salinity maximum zone, (b) the export of salt from mangrove creeks.

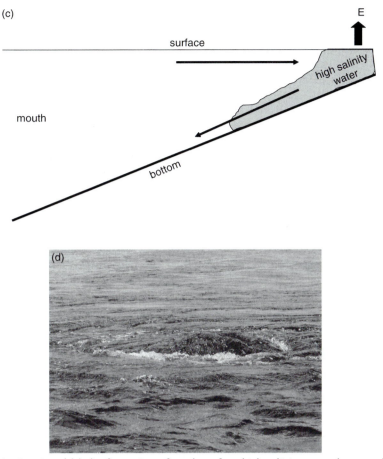

FIGURE 2.8. Continued (c) the formation of a sub-surface high salinity water layer, and (d) a surface boil of water indicating a submarine freshwater spring along the Yucatan peninsula, Mexico. Q=flow rate, E=evaporation, S=salinity. Photo (d) is courtesy of J. I. Euan-Avila.

fringing intertidal wetlands, and this increases water salinity in the creek. Waters become hypersaline. However the salinity S reaches a maximum usually quickly (within 2 weeks) and does not increase thereafter. A steady-state solution thus forms whereby the excess salt is exported by tidal diffusion,

$$A K_x \, dS/dx = e \, W \, S \tag{2.19}$$

where x is the distance along the tidal creek (x = 0 at the mouth), A is the cross-sectional area of the creek, S is the salinity, and K_x is the longitudinal diffusion coefficient. K_x can be computed using Eq. (2.19) because all the other terms in this equation can be measured in the field. The residence time T can then be calculated from the equation for turbulent diffusion scaling (Fischer et al., 1979),

$$T = L^2/K_x \tag{2.20}$$

where L is the length of the tidal creek (estuary). For typical mangrove-fringed tidal creeks, $L \approx 5\,km$, $K_x \approx 30\,m^2\,s^{-1}$, and $T \approx 10$ days (Wolanski et al., 2001).

If tidal mixing is subdued, the high salinity water downwells in the upper reaches the tidal creek and propagates seaward along the bottom (Fig. 2.8c).

If the estuary is vertically stratified, then salinity is not just a tracer, it actually determines the pathways of water, hence the residence time. If the estuary is vertically well-mixed, salinity data can be used to determine the residence time of an estuary. For tropical estuaries in the dry season, Eq. (2.18) is used to estimate K_x, and from there the residence time T can be calculated from Eq. (2.20). For classical, temperate estuaries, K_x can be calculated from salinity field data at steady state (no significant changes in the river discharge Q_f for a period larger than T) from the equation for conservation of salt,

$$-K_x\, A\, dS/dx = Q_f S \qquad (2.21)$$

and from there T can be estimated from Eq. (2.20).

The assumption of a constant value of K_x is very restrictive for many estuaries where A decreases rapidly with increasing value of x (i.e. the distance from the mouth). If A is constant and mixing is tide-driven, then K_x is a constant. A is not a constant if mixing is due to gravitational circulation. In reality both mixing mechanisms occur in different parts of the estuary. The relative importance K of these mechanisms near the toe of the salt intrusion curve can be estimated from the Van den Burgh equation (Savenije, 2005),

$$dK_x/dx = -K\, Q_f/A \qquad (2.22)$$

If $K \ll 1$, mixing is tide-driven; if $K \approx 1$, gravitational mixing is dominant.

Some estuaries are vertically mixed by groundwater inflow (Fig. 2.8d). The inflow of groundwater at the bottom makes the water mass unstable, leading to overturning. The resulting ventilation of the bottom waters prevents anoxic conditions. Mixing by freshwater springs is common in coastal waters of the Yucatan Peninsula – they can bring nutrients (from farm and cities) and pollutants direct to coastal waters, bypassing the estuary. Groundwater intrusion of pollutants commonly occurs along limestone coasts where sewage from septic tanks directly flows into coastal waters and degrades seagrass and coral reefs. Such is the case in the Florida Keys, Zanzibar, the Yucatan peninsula, and the northern Kenya coast.

2.5. LATERAL STRATIFICATION, TRAPPING, AND SHEAR

The water circulation in estuaries can also vary markedly across the width. In wide estuaries, the Coriolis force causes a horizontal shear of the flow; as a result, seaward flow occurs on the right (left) hand side in the northern (southern) hemisphere, and landward flow on the left (right; Dyer, 1997). In wide estuaries, this creates (Fig. 2.9a) a tidally-averaged, net intflow on one side (this is called the flood channel) and a net outflow on the other side (the ebb channel).

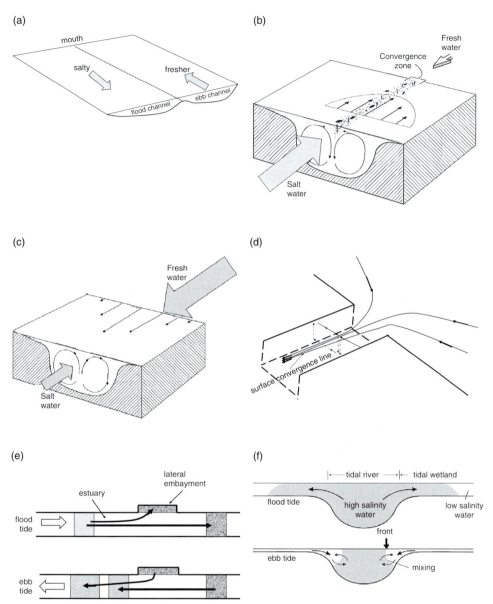

FIGURE 2.9. Sketch of the secondary circulation in estuaries. (a) Tidally averaged flows in a well-mixed estuary with distinct flood and ebb channels over an uneven bottom and when upstream salinities are reduced due to freshwater, (b) a surface convergence is formed at flood tide and (c) a surface divergence is formed at ebb tide, (d) surface convergence due to the converging flows at rising tide near the river mouth, (e) trapping in lateral embayments, including tidal wetlands, (f) cross-channel distribution of salinity in a wetland-fringed estuary at flood and ebb tide. In (f) the arrow points to the location of the oceanographic front forming near the banks at ebb tide. (a) is redrawn from Dyer (1997), (e) is redrawn from Okubo (1973), (f) is modified from Wolanski (1992).

Across an estuary the tidal currents are generally larger in deeper than in shallow waters, because of friction. At rising tide, the incoming, high salinity waters travels faster in the deeper regions of the estuary (Fig. 2.9b) and brings higher salinity ocean water. This water is denser than the surrounding brackish estuarine water on the sides. It sinks and spreads laterally, entraining water from near the banks to form a mid-channel axial convergence (Simpson and Turrell, 1986). Floating material, such as plant litter from the fringing wetlands, is aggregated along the convergence line. At falling tide, the process is reversed; axial convergence occurs near the bottom and axial divergence prevails at the surface. Floating matter is pushed towards the banks (Fig. 2.9c).

Axial convergence at the surface, and downwelling along the convergence line, also occurs in a constriction (Fig. 2.9d; Wolanski and Hamner, 1988).

There can also be considerable cross-channel, three-dimensional currents in meanders and around shoals (Friedrichs and Hamrick, 1996; Dyer, 1997).

Lateral trapping occurs in the presence of lateral embayments, a process first studied by Okubo (1973). A fraction of the water moving upstream at flood (rising) tide moves into the lateral embayment. At ebb (falling) tide it returns to the estuary and mixes with untagged water. This process enhances longitudinal mixing (Fig. 2.9e). This is expressed by a enhanced value of the tide-averaged longitudinal diffusivity B from its original value K_z,

$$B = K_z /(1+\varepsilon) + \varepsilon U_{max}^2 /2k(1+\varepsilon)^2(1+\varepsilon+\sigma/k) \qquad (2.23)$$

where ε is the ratio of the volume in the lateral embayment to that in the estuary, U_{max} is the peak tidal current, $1/k$ is the characteristic exchange time between the embayment and the estuary, and σ is the tidal frequency. For estuaries fringed by intertidal wetlands, such as saltmarsh, mangroves, mud banks and salt pans, Eq. (2.21) is modified to account for the fact that the depth of water in the estuary and the wetlands are not the same, so that the exchange of water between the estuary and the creek is not constant in time but occurs during a small fraction of the tidal cycle,

$$B = K_z/(1+\varepsilon) + \pi \varepsilon U_{max}^2 a^2 / 48\sigma(1+\varepsilon) \qquad (2.24)$$

where a is the fraction for the time that the wetland is inundated by the tides.

Longitudinal mixing can be very intense (typically B $\approx$ 10–40 m^2 s^{-1}) in wetland-fringed tidal creeks, and is much smaller without wetlands (typically B $\approx$ 1–10 m^2 s^{-1}) (Wolanski and Ridd, 1986).

In the presence of buoyancy, mixing may be inhibited when freshest water is pushed back into the tidal wetlands at flood tide and is isolated from the tidal currents in the main estuary (Fig. 2.9f). This water forms a brackish water plume tagging the banks at ebb tide, made evident by a small-scale oceanographic front at the edge of the plume and parallel to the banks.

Moderate shoreward wind (speed 5–10 m s^{-1}) generates small breaking waves on steep shores. Floating material is aggregated along a slick line parallel to the coast and located typically 2–20 m away from the coast (Fig. 2.10). This floating material is not washed on the shore, although the wind could do that in a few seconds. The breaking waves create a water set-up that generates a seaward

FIGURE 2.10. Vertical photograph of waves breaking on a rocky shore, with a shoreward wind of $5\,\mathrm{m\,s^{-1}}$. The arrows point to a slick line of floating material that is not washed ashore because the seaward surface current generated by the water step-up from breaking waves counteracts the shoreward wind.

surface current that is opposite to the shoreward wind. These opposite effects meet at the convergence point where they form a slick line parallel to the shore.

2.6. THE IMPORTANCE OF THE BATHYMETRY ON FLUSHING

The rising tide propagates landward along the length of the estuary. On reaching the head of the estuary it is reflected and returns seaward. The estuary is tidally resonant if the time to do that equals the tidal period; in that case the exiting wave meets the next tidal wave entering the estuary from the sea. A standing wave forms resulting in the tidal range at the head of the estuary being twice that of the original wave. At a distance of one quarter that of the wave length of the wave, an antinode exists with zero tidal variation of the water depth. In long estuaries several nodes and antinodes may exist. In short estuaries, high and low waters occur simultaneously throughout the estuary. In very shallow estuaries, bottom friction dissipates the energy of the tidal wave that becomes solely a progressive wave, the tidal range of the peak tidal currents diminishing with distance from the mouth. Many estuaries fall in between these two types, i.e. the tide is a combination of a progressive wave and a standing wave; the current continues flooding (to flow landward) until after the water level starts to recede (Pendleton and FitzGerald, 2005).

In an unstratified estuary the tidal wave travels at a speed c,

$$c = (g\,h)^{1/2} \tag{2.25}$$

where g is the acceleration due to gravity and h is the depth. Thus, the wave travels faster over deep water than over shallow water. This leads to a tidal asymmetry whereby the peak current is larger at ebb tide (ebb dominance) or

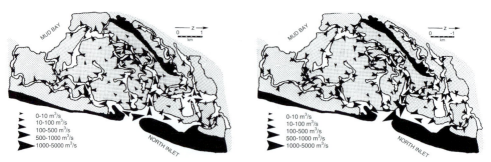

FIGURE 2.11. Predicted flow field in the North Inlet Estuary and salt marsh system, South Carolina, USA, at (a) flood tide and (b) ebb tide. Adapted from Kjerfve et al. (1991).

at flood tide (flood dominance). Whether the estuary is flood dominant or ebb dominant depends on the bathymetry, the river inflow, and the extent of intertidal wetlands. Saltmarsh creeks and mangrove creeks are ebb dominant, i.e. during spring tides the peak ebb tidal current at the mouth of the creek is measurably larger than the peak flood tidal currents (Fig. 2.11; Kjerfve et al., 1991; Wolanski, 1992; Wolanski et al., 2001).

The curvature of the meanders generates complex 3-dimensional flows. In vertically well-mixed waters and smooth meanders (i.e. with a large radius of curvature), the flow forms a helix, with upwelling along the inner bank and downwelling along the outer bank (Fig. 2.12a; Dermuren and Rodi, 1986). In a highly stratified estuary, the density interface slopes upward toward the inner bank and the upwelling and downwelling are largely confined to the bottom layer (Fig. 2.12b; Wolanski, 1992). In a cuspate meander, i.e. a meander with a small radius of curvature, an eddy forms downstream, and a smaller eddy may occur upstream on the opposite bank (Fig. 2.12b). These transient eddies behave as lateral embayments and enhance longitudinal mixing following Eq. (2.23).

Complex flows, including eddies, are shed by headlands (Fig. 2.13), islands (Fig. 2.14), shoals, and man-made structures such as groynes along the banks of estuaries. The flow field in such eddies is unsteady. The flow field in the lee of obstacles can take many forms, sketched in Fig. 2.15, according to the value of the island wake parameter P,

$$P = U \, W/K_z \, (H/W)^2 \qquad (2.26)$$

where K_z is the vertical eddy diffusion coefficient. U is the flow velocity meeting the obstacle, W is the width of the island (the length of the headland or the groyne), and H is the depth (Wolanski et al., 1984; Wolanski et al., 1996a). For $P < 1$, the flow does not separate and there is no eddy. As shown in Fig. 2.15, for $P \approx 1$, an eddy or an eddy pair exists (case a in Fig. 2.15; Fig. 2.14a); for $P = 1-3$, meanders develop (case b in Fig. 2.15; Fig. 2.14b); for $P = 3-15$, the meanders develop instabilities and roll (case c in Fig. 2.15; Fig. 2.14c); for $P > 20$, the wake is fully turbulent downstream (case d in Fig. 2.15; Fig. 2.14d). The flows in such eddies in shallow water is three-dimensional, with an upwelling in the eddy's centre and a downwelling along the sides of the eddy. This downwelling is made

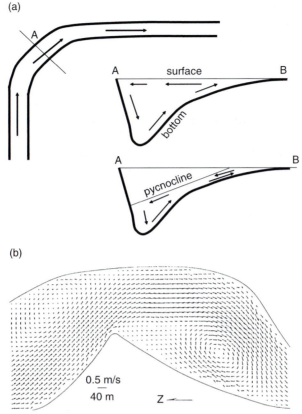

FIGURE 2.12. Synoptic distribution of the water currents (a) around a smooth meander for (top) well-mixed and (bottom) stratified waters, and (b) around a cuspate meander of the South Alligator Estuary, Australia, at flood tide. (b) is modified from King and Wolanski (1996).

FIGURE 2.13. Aerial photograph of the counter-clockwise rotating eddy shed by a headland in shallow coastal waters, Whitsundays, Australia. The arrow points to the convergence zone along the edges of the eddy, made visible by the foam line.

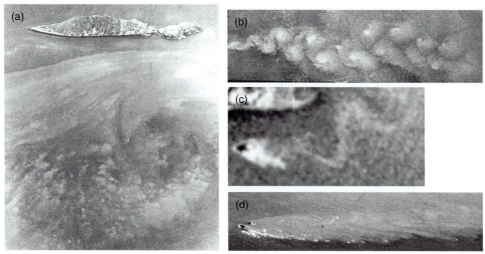

FIGURE 2.14. Aerial photographs of eddies behind islands in shallow coastal waters. (a) the 1500 m wide Rattray Island, Australia, (b) small (10–100 m wide) islands in the Whitsundays and Torres Strait, Australia.

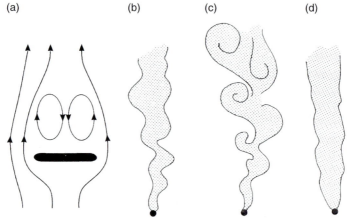

FIGURE 2.15. Sketch of the shallow water island wake for different values of the island wake parameter P. (a) P $\approx$ 1, (b) P $\approx$ 1–3, (c) P $\approx$ 3–15, (d) P > 20.

visible by the formation of a foam layer along the edges of the eddy (see the arrow in Fig. 2.13). The upwelling lifts fine sediment to the surface and makes the eddy turbid and visible (Fig. 2.13).

For the tidal flow past an embayment, one or two eddies may exist in the embayment depending on the length of the embayment (Fig. 2.16; Uijttewall and Booij, 2000; Valle-Levinson and Moraga-Opazo, 2006). A shear layer separates the fast flowing tidal currents in the main body of the estuary from the eddy, or eddies, in the embayments. This free shear layer is increasingly sharp with increasing speed of the prevailing flow in the estuary (Fig. 2.17a and b). When the

(a)   (b)

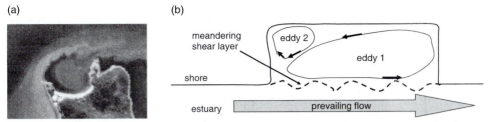

FIGURE 2.16. (a) Aerial photograph of a solitary eddy in a small embayment, Cape Richards, Australia. (b) Sketch of the eddy pair and the free shear layer that can meander and form rolling vortices in a long embayment.

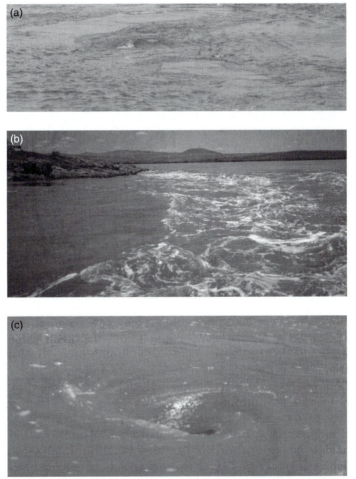

FIGURE 2.17. Photographs of a smooth shear layer for a prevailing flow approximately equal to (a) $0.5\,\mathrm{m\,s^{-1}}$ at Bench Point, Australia, (b) $2\,\mathrm{m\,s^{-1}}$ in Cambridge Gulf, Australia. (c) A photograph spanning about 1m of an intense eddy in the shear layer shown in (b). In (a), the smooth areas indicate recently upwelled water, the surface of which has not yet been disturbed by waves.

shear layer is steady, the water in the embayment is, in practice, nearly completely trapped. The free shear layer has energetic eddies imbedded within it, with very large vertical velocities (up to $0.2 \, m \, s^{-1}$; Fig. 2.17c; Wolanski, 1994). It also develops instabilities such as meanders that develop into rolling vortices, this promotes mixing between the waters in the main body of the estuary and those inside the embayment (Uijttewaal and Booij, 2000). The main impact of these processes on the estuary is to enhance the longitudinal diffusion coefficient, following Eqs. (2.23) and (2.24). If pollution is a problem in the estuary, the pollution impact may be exacerbated in these embayments.

2.7. THE IMPORTANCE OF FLOWS NEAR THE RIVER MOUTH ON FLUSHING

The shape, depth and width of the river mouth determine the return coefficient, i.e. the fraction of the estuarine waters that leave the estuary at falling tide and to the estuary at rising tide. This is a dominant process controlling the flushing of an estuary.

Unobstructed, wide and deep river mouths enable tidal exchange and mixing between estuarine and coastal waters. By buoyancy effects, the outflow forms a river plume, and the internal circulation within the plume aggregates floating matter, including plankton and detritus, along the river plume front, forming a distinct foam line (Fig. 2.18a and b). Whether this water will or will not return to the estuary depend on the coastal circulation.

If the coastal currents are small and the river mouth is large, the river plume does not form a jet at falling tide; instead it forms a radially symmetric plume (Fig. 2.19a). This will increase the return coefficient because much of that water will return into the estuary at flood tide since flood tidal currents will also be radially symmetric if the coastal currents are small (Fig. 2.20c and d discussed below). If the coastal currents do not reverse direction with the tides, the plume never returns to the estuary and the return coefficient is the smallest it can be,

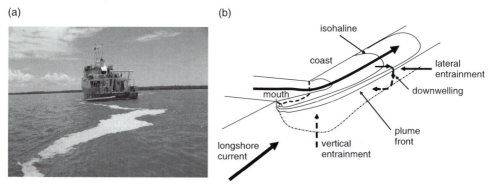

FIGURE 2.18. (a) A photograph of a foam line indicating a river plume, Darwin Harbour, Australia. (b) A sketch of the water circulation in a river plume.

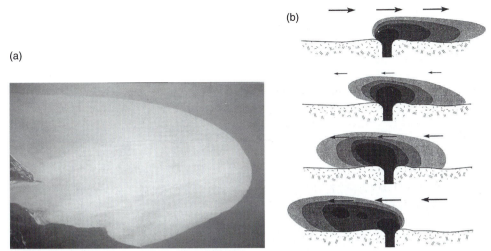

FIGURE 2.19. (a) Aerial photograph of the radially symmetric Moresby River plume, Australia. (b) Sketch of the salinity distribution at various stages of a river plume in a reversing coastal tidal current, leading to patchiness. Darker shading indicates fresher water. Before tidal flow reversal the low-salinity (black) is to the right of the river mouth, As the flow reverses toward the left, the 'new' river water mixes with the 'old' plume water (stippled) and the low-salinity patch grows, When the reversing current is fully established and the low-salinity patch is deflected to the left, low-salinity patches are imbedded in the river plume. During the flow reversal process, 'old' estuarine waters return into the estuary at flood tide, increasing the return coefficient. (b) is adapted from Wolanski et al. (1999).

approaching zero. If the coastal currents reverse direction with the tides, the return coefficient is increased because some estuarine waters can return in the estuary from offshore (see Fig. 2.19b). The flow reversal causes the plume to have patches of low-salinity waters imbedded within higher salinity plume waters.

When the mouth is narrow, the tidal flow in coastal waters off the river mouth takes the form of a tidal jet at falling tide and of a sink flow at rising tide (Stommel and Farmer, 1952; Wolanski et al.,1988a; Wells and van Heijst, 2003). When the tidal jet is unstable, the estuarine water is rapidly mixed in coastal waters (Fig. 2.20a) and some of that water is readily re-entrained back in the estuary at rising tide. When the tidal jet is stable (Fig. 2.20b), the outflow at falling tide comprises a jet and a dipole (a vortex pair) at its leading edge. The return coefficient of the estuary depends on whether the dipole is re-entrained into the estuary when the tide reverses. When the tide reverses, this dipole may continue to move away from the mouth or it can be re-entrained into the estuary by the sink flow, according to whether the self-propelled dipole velocity V_{dipole} at ebb tide is larger or smaller than the velocity V_{sink} at flood tide. In practice this is determined by the value of W/UT, where W is the width of the mouth, T the tidal period, and U the jet velocity (Wells and van Heijst, 2003). If W/UT is less than 0.13, the dipole propagates away from the mouth without being re-entrained into the estuary at flood tide. This decreases the return coefficient, and thus increases the flushing of the estuary. If the dipole is re-entrained into the

(a)

(b)

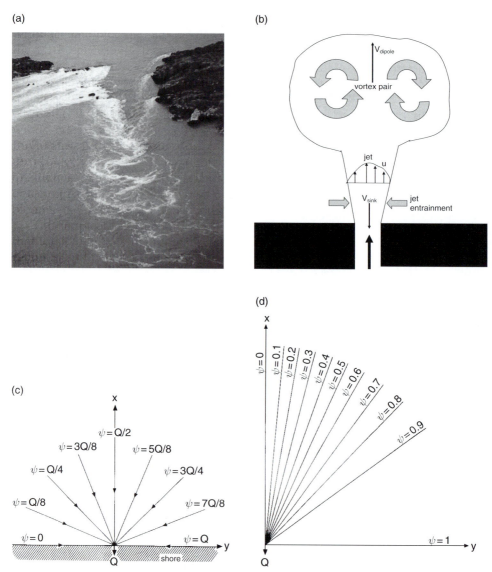

(c)

(d)

FIGURE 2.20. The ebb tide through a narrow mouth induces a tidal jet that can be (a) unstable or (b) stable. The stable tidal jet grows in width by jet entrainment, resulting in maximum velocity u near the jet axis, and pushes forward a dipole (a vortex pair). At flood tide the streamfunctions ψ marked as a fraction of the tidal discharge Q, show (c) radially symmetric for a flat bottom, and (d) a funnel flow for a sloping surface in shallow water. (c) and (d) are adapted from Wolanski and Imberger (1987).

estuary, the value of the return coefficient is increased. The degree to which this happens depends on the bathymetry (Wolanski and Imberger, 1987). For a flat sea floor, the flow at rising tide is radially symmetric (Fig. 2.20c) and the return coefficient is minimized. For a sloping seafloor in shallow waters, the flow at rising tide is funnel-like with the streamlines concentrated offshore (Fig. 2.20d) and therefore the return coefficient increases.

2.8. THE SPECIAL CASE OF LAGOONS

Section 2.7 demonstrates the importance for flushing of the dynamics near the mouth. This is particularly important for lagoons, which are shallow estuaries chocked at the mouth. They often are elongated parallel to the ocean and separated from the ocean by a series of barrier islands. Inlets, either natural or man-made, cut through barrier islands and permit flushing with the ocean. The mouths are narrow and shallow and flushing is thus restricted. This generates three important characteristics of lagoons. (1) The mouth measurably inhibits the propagation of the tides to the inner parts of the system. (2) Flushing is restricted and extensive flooding of adjacent lands can occur during the wet season. (3) The residence time is large, typically a few months.

Lagoons have been subdivided into "choked", "restricted" and "leaky" systems, an arbitrary but useful terminology indicating a decreasing residence time (Kjerfve, 1986; van de Kreeke, 1988). Accordingly, a choked lagoon may have no more than one inlet, a restricted lagoon has a wider water body and two or more inlets connecting the lagoon and the sea, and a leaky lagoon has many entrances among the barriers. The same lagoon can be choked in the wet season and restricted in the dry season. Thus in the wet season it may have water level rising by typically 1 m and the water becoming fresh or brackish, with a residence time of typically several months. It can experience strong seiching (set-up/set-down cycles) related to wind forcing of typically 0.2–0.5 m. As a restricted lagoon in the dry season tides may be readily transmitted but attenuated into the lagoon.

Lagoon waters are usually well mixed vertically. Salinity fluctuates less in a restricted lagoon (typically 10–35) than in a choked lagoon (typically 1–80).

A formal lagoon classification relies on the 'repletion coefficient' K (Keulegan, 1967; van de Kreeke, 1988),

$$K = \tau_t (A_c/2\pi a_o A_b)\{2g\,R\,a_o/[(\alpha+\beta+\gamma)R+2f\,L_c]\} \tag{2.27}$$

where τ is the tidal period (e.g. 12.42 h for the M_2 tide), a_o is the tidal amplitude (i.e. half the tidal range), A_c is the inlet cross-section area, A_b is the lagoon surface area, R is the hydraulic radius at the mouth ($\approx$ depth if the mouth is much wider than it is deep), f ($\approx 116\,n^2\,R^{-1/3}$, where n is the manning friction coefficient) is the Darcy-Weisbach bottom friction factor, L_c is the length of the inlet, α is a coefficient describing the vertical velocity distribution (0.1–0.3), β is the fraction of tidal kinetic energy dissipated in the transition between the entrance and the inlet channel and γ is the fraction of remaining kinetic energy that is lost at the lagoon side of the inlet. Values of K < 0.3, 0.3–0.8 and > 0.8 define the three lagoon types.

CHAPTER **3**

Estuarine sediment dynamics

3.1. GEOMORPHOLOGICAL TIME SCALES

The mean sea level (MSL) as measured from coastal stations was 120–130 m lower 20,000 years ago than at present (Fig. 3.1a; Pirazzoli, 1991). Most present estuaries were river valleys discharging water in estuaries located where the present continental slope is. Because the continental slope is steeper than present coastal plains, the estuaries were generally short. The MSL rose quickly and reached the mouth of the present estuaries about 10,000–12,000 years ago, when the present estuaries started to form. There was a marked slow-down of the MSL rise between 7,500 and 8,000 years ago shortly after the cessation of melt-water input to the oceans from the northern hemisphere. The maximum MSL was reached about 6,500 years ago (Fig. 3.1b). The MSL decreased thereafter by about 3 m due to hydrostatic rebound of the continental plates relieved of the pressure of ice glaciers. The MSL has been fairly constant during the last 6,000 years. It is now at a high stand level that was last reached about 120,000 years ago.

Estuaries responded to this MSL change by infilling with sediment derived from both the riverine inflow and from the sea (Fig. 3.2; Chappell, 1993; Perillo, 1996; Woodroffe, 2003). The estuaries are also adjusting to human-induced changes in river discharge Q_f and riverine sediment inflow Q_s. Most estuaries have not reached a steady state yet; they are still evolving. They have an age; they all started young; most have reached maturity though most fjords still have not; and some have reached old age meaning that they cannot accommodate any additional sediment. The extra sediment is carried seaward to be dispersed at sea or to form a delta. This often involves the lateral displacement of the river mouth following a major river flood.

When river valleys were drowned by risings seas, inland seas were formed (Chappell, 1993; Woodroffe, 2003). These were settling basins that infilled with sediment from both the river and the sea. The latter sediment was brought in from the sea by waves, longshore currents, and tides. The estuaries formed in different ways, largely determined by the topography when the sea level rose at

41

(a)

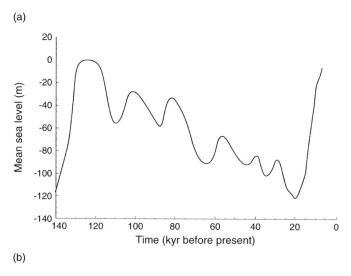

(b)

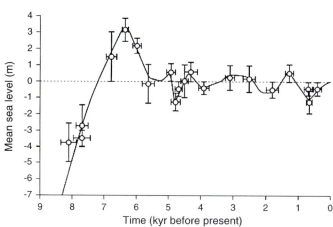

FIGURE 3.1. Time series plot of the mean seal level (MSL) over (a) the last 140,000 years and (b) the last 8,000 years. The MSL is a global average; some estuaries are still locally sinking or rising relative to MSL due to local tectonic movements. (a) is redrawn from Johnston and Lambeck (1999), (b) from Lambeck and Chappell (2001).

the end of the last glaciation. The sediment infilling models include (Woodroffe, 2003),

- *The big-swamp model*
  This usually applies to macrotidal estuaries. A large swamp (Fig. 3.3a) formed, colonized by mangroves (in the tropics) and saltmarshes (in temperate regions). The swamp kept up with sea level rise by importing mud from the sea at a rate $Q_{s,in} > Q_s$. Ultimately freshwater sediment from rivers capped the saltwater sediment, at which stage the estuary reached old age.

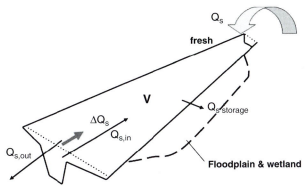

FIGURE 3.2. The infilling of an estuary of volume V is controlled by the inflow of riverine sediment Q_s, the net balance ΔQ_s between seaward ocean flux $Q_{s,out}$ and the landward sediment flux from coastal water $Q_{s,in}$, and the sediment flux to lateral storage in the flood plains $Q_{s,storage}$.

- *The progradation model*
  A coastal mangrove or saltmarsh swamp formed fringing the coast (Fig. 3.3b). It migrated upwards with sea level rise and left its mud behind, submerged by rising seas and later capped by marine sand. When sea level stabilised, the swamp prograded seaward and was progressively capped by freshwater (river) sediment. This system reached old age when reaching offshore where the steep bed slope and the waves prevented growth.
- *The basin-lagoon model*
  Waves built up a barrier of sand and prevented much sediment inflow from the sea ($Q_s > Q_{s,in}$; Fig. 3.3c). The lagoon was a sink for fluvial sediment. Mud was deposited in the deeper portions and also near the coast where swamps formed. The mud basin progressively filled. The estuary reached old age when the basin was full and the river discharged its sediment directly into the sea.

The accepted method to quantify such sediment infillings is through analyzing sediment cores. The analysis relies on radionuclide dating, including for recent history the short-lived anthropogenic (atomic bomb tests during the 1950s) radionuclides, and inferring past habitats through pollen analysis (Brush et al., 1982; Woodroffe et al., 1986; Fletcher et al., 1993). Following nuclear testing during the 1950's and 60's, ^{239,240}Pu, with a half life of thousands of years, and ^{137}Cs, with a half life of 30 years, were distributed around the globe. Pu binds strongly to soil particles and remains tightly bound to soil particles upon entering saline waters, thus it is well suited as a tracer of soil transport. ^{137}Cs tends to desorb upon entering saline waters. Radionuclide dating is well suited to unchanging sites such as backwater lakes on a floodplain connected to the estuary by a channel). This technique is not reliable in a geomorphologically dynamic, meandering estuary (Fig. 3.4a and b). The moving meanders erode 'old' sediment from the eroding bank and lock 'new' sediment in the accreting bank. Sediment cores then yield a mixture of 'old' and 'new' sediment. For such estuaries long-term

(a)

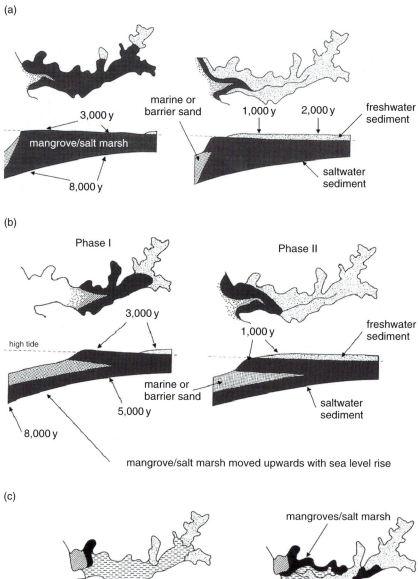

(b)

(c)

FIGURE 3.3. Most estuaries have followed one of the three following evolution models when infilling with sediment since the last glaciation. (a) the big swamp model, (b) the progradation model, (c) the basin-lagoon model. The numbers are years before present. Modified from Woodroffe (2003).

(a)

(b)

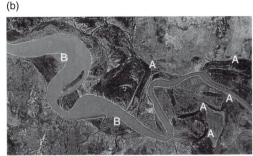

FIGURE 3.4. (a) Photograph of the banks of the Fly River Estuary, Papua New Guinea. Bank erosion is so rapid that still living palm trees are left standing in the estuary. (b) Satellite photograph of a 30 km-long stretch of the Daly Estuary, Australia, where meanders migrate through the estuarine flood plains at a rate of up to 25 m year^{-1} (Chappell, 1993). In so doing, they erode 'old' sediment from the eroding bank and deposit 'new' sediment on the accreting bank. This results in mixing 'new' and 'old' sediment. Remnants of meanders (A) and accretion zones (B) are highlighted.

estimates of erosion/sediment yield can be obtained using cosmogenic ^{10}Be in quartzose rock from selected ground-surfaces in the catchment, coupled with ^{10}Be in sedimentary quartz grains from the estuary.

3.2. SEDIMENT DYNAMICS

3.2.1. The distinction between mud, silt and sand

Cohesive sediment (also called mud or clay) have a mean particle size d_{50} (also called d_s) $< 4\,\mu$m (microns) (1 mm $= 1000\,\mu$m; Postma, 1967). Non-cohesive sediment (sand) has a $d_{50} > 64\,\mu$m. Silt has a d_{50} in between those of mud and sand; this is the hardest sediment to study because it is weakly cohesive.

The sediment dynamics depend on the currents, the salinity, and d_{50}. For mud, they also depend on the biology, primarily the plankton in the water and the bacteria living on the mud particles in suspension and on the bottom.

Mud and sand affect the environment in different ways. They are transported in the water column in different ways. Sand is mainly carried along the bottom and very close to the bottom. Mud is mainly carried in suspension within the water column. Both sand and mud transport affect the bathymetry by sedimentation and erosion. Both sand and mud transport negatively affect the benthos both by destabilizing the substrate and by burying the benthos. Mud also affects the biology of the water because it increases the turbidity; less light generally means less photosynthesis. Light penetration depends on the coloured dissolved organic material (CDOM) and the dissolved suspended solid concentration (SSC). CDOM, or yellow substances as they are sometimes referred to, results from the rotting of organic detritus that releases tannins into the water, staining the water yellow to brown. Organic detritus includes decaying phytoplankton and plant matter such as leaves.

At high SSC (highly turbid), there is nearly complete darkness. Typically, for $SSC = 0.2\,g\,l^{-1}$ ($200\,mg\,l^{-1}$) , visibility $< 30\,cm$; for $SSC = 1\,g\,l^{-1}$ ($1000\,mg\,l^{-1}$), visibility $< 1\,cm$. A plankton in turbid waters can only photosyntesise for the short time that it is brought to the surface by turbulence. The benthos has no such opportunity; it stays in complete darkness as long as the water stays turbid.

Even at small values of suspended matter concentration ($< 4\,mg\,l^{-1}$, a value often associated with 'clear' water), light (i.e. the visibility) is still strongly atten-uated (Fig. 3.5; Binding et al., 2005; Bowers and Binding, 2006). Suspended matter consists of suspended sediment and biological matter.

Even in the absence of suspended sediments, light is attenuated with depth. Different light attenuation curves prevail for different environments depending on the plankton (shading effect) and on dissolved organic matter (DOM). The coloured part of DOM is called Gelbstoff or yellow substance and is an important factor in light attenuation (Bowers et al, 2000). To predict the light absorption, scattering and backscattering requires extensive measurements of chlorophyll, total suspended solids and carbon dissolved organic matter. The relationship can vary markedly from place to place (Bowers et al., 2004; McKee and Cunningham, 2006).

Mud also affects heavy metals and nutrients because these particles are readily absorbed on the mud due to the electric and colloidal forces on the clay particles. Thus pollutants can stay trapped hundreds of years on settled mud and are released when the mud is resuspended (Fig. 3.6). This generates a serious problem particularly for European and U.S. estuaries in deciding what to do with harbour sediment polluted by heavy metals sometimes one or two centuries ago.

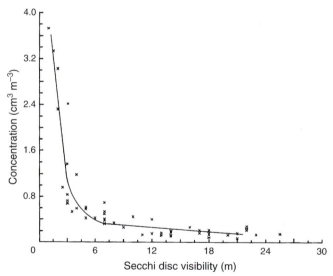

FIGURE 3.5. Secchi disc visibility as a function of suspended matter concentration for Great Barrier Reef coastal waters. Adapted from Wolanski et al. (1981).

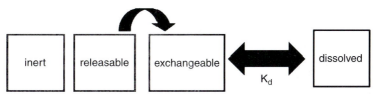

FIGURE 3.6. Heavy metals partitioning in muddy estuaries. A somewhat similar scheme also holds for some nutrients, particularly phosphorus.

The partition coefficient K_d for an element is defined as (Salomons and Forstner, 1984),

$$K_d = \text{mass absorbed on particulates/mass in solution} \qquad (3.1)$$

Its value is different for different heavy metals (e.g. Cd, Pb, and Cu) and varies with salinity, pH, and the amount of organic matter on the sediment (Millward and Liu, 2003). Thus, because the metal is distributed over different phases, a simple measurement of total concentrations of the metal is inadequate to assess its bioavailability. It is necessary to measure the metal or the nutrient in the various phases and this is usually done using sequential extraction, e.g. using increasingly stronger acids (Perin et al., 1997). Higher salinity liberates more metals into the dissolved form. Higher organic content of the suspended sediment helps reduce bioavailability.

Partitioning between dissolved and particulate phases also applies to nutrients. Phosphorus is much more readily absorbed on suspended mud than nitrogen. Neither nitrogen nor phosphorus absorbs much on non-cohesive sediment (sand). Thus the impact of riverine nutrients on an estuary is very different for a sandy (clear water) estuary where the nutrient is readily bioavailable and light penetration allows photosynthesis, than for a muddy (turbid) estuary where much of the nutrient, especially phosphorus, is absorbed on the mud and photosynthesis is inhibited by the lack of light.

3.2.2. Sand dynamics

There are many engineering models to calculate the bed load; these include the Meyer-Peter, Einstein, and Ackers-White formulae (see a review in Raudkivi, 1967; Postma, 1967; Dyer, 1986 and 1994, Camenen and Larson, 2005; Chanson, 1999). The models generally assume either that the sand moves as a creeping movement of layers along the bottom (Fig. 3.7a) or that the sand particles are rolled and tossed over the bottom in a process called saltation (Fig. 3.7b). They propose empirical formulae, largely from laboratory studies, to estimate the bed load transport as a function of the current-induced stress on the bottom sediment, a threshold stress for bed load to commence, the difference in density between the water and the sediment, and d_{50}. They are reliable in rivers for steady flows. They are much less reliable in estuaries because the sediment transport lags behind the currents when the currents vary in time with the tides. They are even less reliable for the coastal zone because of the additional effect of waves that resuspend

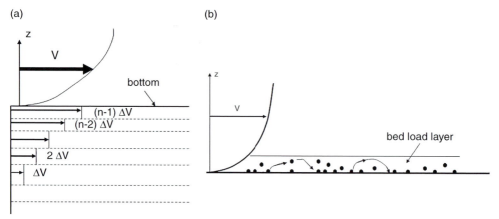

FIGURE 3.7. Models of bed load transport by (a) creep flow below the bottom and (b) saltation above the bottom.

sediment. These formulae yield bed load predictions that can vary by a factor of 10 with each other and from field measurements.

A reason for these discrepancies is the multiple effects of bedforms, which include sand dunes and ripples. It is in practice not possible to capture all these effects within one formula. Instead empirical coefficients are often used to adapt a bed load transport formula to local field data. The effects of bedforms include increased bottom roughness length, changing the flows and wave propagation in shallow waters, and patchiness of sediment transport especially near crests of sand waves and ripples (Dyer, 1986; Ke *et al.*, 1994; Le Hir *et al.*, 2000; Dronkers, 2005).

The bed load transport is generally assumed to increase as the 4th power of the water velocity, with sand transport commencing only when the water velocity exceeds a threshold value u_c (typically about 0.3 m s^{-1}). Therefore for usual currents in the range 0.5–1 m s^{-1}, the erosion rate can be up to 8 times larger if the velocity doubles. The erosion rate is thus very sensitive to large currents even if they last a short time. This is important in many shallow estuaries where a tidal asymmetry prevails, i.e. where the flood tide is shorter in time than the ebb tide. This results in higher flood than ebb tidal currents, though the flood currents last a shorter time; as a result the net transport of sediment is landward. This is the process called tidal pumping.

Sand beds are rarely pure in estuaries. Sand is often mixed with mud even if the sand is the dominant fraction. The bed load transport can be inhibited, or enhanced, by the presence of mud particles within the sand. This depends on whether the mud particles are found in sufficiently small quantities to lubricate the sand by preventing sand particles to interlock (Barry et al., 2006), or in sufficiently large quantities to cement the sand particles (Wolanski et al., 2006a). From limited data it appears that this effect of clay particles is significant enough to modify the bed load by a factor of 2 to 4. This requires modifying the erosion constants in the bed load transport equations, for which extensive field data are needed and are still lacking.

Understanding and modeling the stability of sandy coasts attacked by waves, a process of great interest to coastal engineers, remains an art more than an exact science. It is the usual practice to divide the coastal area into three zones. These are (1) the sediment-active zone extending from the beach to the storm wave breaking zone; (2) the active zone for sediment dispersion that extends from the storm wave breaking zone to the zone where wave shoaling becomes significant; and (3) a non-active zone further offshore. The storm wave breaking zone is the location of the outermost offshore bar. The shape of the bed can be calculated depending on the wave dynamics. It is rarely at steady state; commonly the sandy bed is in a recovery mode following erosion in a severe storm.

One reason for the difficulty in predicting sand transport is that there are no instruments to non-intrusively measure the bed load transport in estuaries. Instead the net bed load has to be determined indirectly by methods such as grain size analysis, magnetism, and repeated bathymetric surveys to assess changes in sediment storage, and the direction of sand transport has to be inferred from the shape of sand shoals and asymmetries in bed forms (FitzGerald and Knight, 2005). Recently, Wolanski et al. (2006a) proposed a new technique to measure the surface velocity of the bed load using a moored, downward-looking acoustic Doppler current profiler. There is still no method, however, to measure the thickness of the bed load layer, but sensitive dual-frequency (30 and 210 kHZ) echo-sounders can provide a rough estimate.

3.2.3. Mud dynamics

Cohesive sediment (mud) behaves differently than non-cohesive sediment (sand). It forms flocs. A floc is typically 50–1000 μm (micron) in diameter, i.e. it comprises thousands of mud (clay) particles (each $< 4\mu$m). The clay platelets are negatively charged on the surface; the dissolved salt ions Na^+ in salt water, and Mg^+ and K^+ in freshwater, form electric bonds between the clay platelet (Fig. 3.8a). The salt neutralizes the negative electrical charges that would cause clay particles to repel each other. As a result the clay platelets move closer together.

Marine snow consists of flocs that comprise a mucus membrane called TEP (transparent exopolymer particles; polysaccharides predominantly exuded by bacteria and diatoms) and dead plankton, fecal pellets and macroscopic aggregates of biological origin Ayukai and Wolanski, 1997; Fabricius and Wolanski, 2000; Wolanski et al., 1998b and 2003 a and b). TEP is very sticky as it is a glue for small mud flocs to aggregate into very large flocs that are called muddy marine snow flocs. In very turbid waters (SSC > 500 mg l^{-1}), marine snow is scarce due to the lack of light for photosynthesis; as a result the flocs are largely inorganic and generally small (Fig. 3.8b). In less turbid waters, TEP is common and the resulting muddy marine snow flocs are very large (Fig. 3.8 c–g). Muddy marine snow flocs are very porous; the floc size generally increases with increasing availability of TEP (Fig. 3.8 b–d). TEP is not just based on mucus; it can also be made of colonies (often several hundreds of μm in diameter) of dead plankton, fecal pellets and macroscopic aggregates of biological origin; small mud flocs aggregate by sticking on this dead organic matter (Figs. 3.8 e–f). In other muddy marine snow flocs,

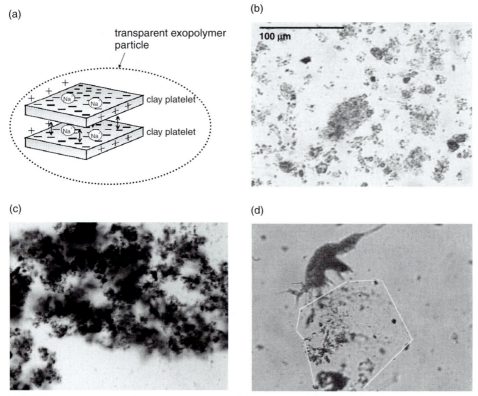

(a)

transparent exopolymer particle

clay platelet

clay platelet

(b)

100 μm

(c)

(d)

FIGURE 3.8. (a) Salt flocculation consists of Na^+ ions helping through electric forces the negatively-charged clay platelets to move closer together. This floc can be enveloped or further held together by transparent exopolymer particles (TEP). (b)–(g) are in-situ microphotographs of flocs. (b) A small floc with little TEP. (c)–(d) In TEP-rich waters, large aggregates form as a result of small mud flocs aggregating to mucus to form large flocs. The aggregation can be dense (c) or light (d). In (d) the floc's outline is shown by the white line. The black dots inside the white outline are small mud flocs glued to the mucus. Such flocs are very porous as is made apparent in this photograph by the fact that the light shines through the floc. A copepod is feeding on the mucus.

clay and silt are segregated in small sub-units that aggregate to form a large floc (Fig. 3.8g). Small mud flocs aggregating on TEP act as a ballast sinking the aggregate, thus forming a biological filter that inhibits the export of mud from estuaries (see Fig. 3.12 discussed later).

The physical processes of erosion and settling have been extensively studied by engineers in laboratory experiments that ignore the biology. The erosion rate E and the settling rate D are parameterised by non-linear laws as a function of the water velocity u (Partheniades, 1965),

$$E = 0 \text{ (no erosion), if } u < u_c \tag{3.2}$$

$$E = A \, (u/u_c - 1)^n, \text{ if } u > u_c \tag{3.3}$$

(e)   (f)

(g)

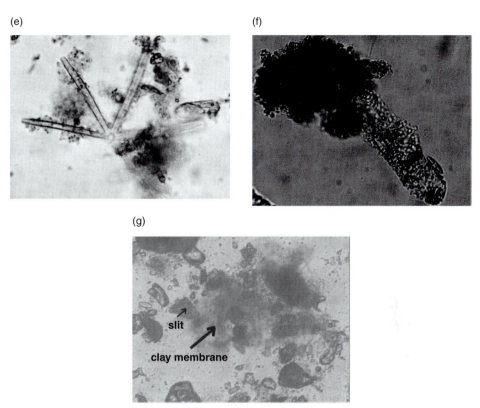

FIGURE 3.8. Continued (e)–(f) Muddy marine snow flocs formed by small mud flocs attached on sticky, dead, organic matter from plankton and fecal pellets. (g) A large muddy marine snow floc formed by mud forming a membrane around large silt particles. The fuzzy area is the clay aggregated in mucus and wrapped around large silt particles. The silt particles are distinguishable from the mud because they have sharp edges and they are black because they are not porous and block the light from the microscope. The width of (c)–(g) is 500 μm.

$$D = C \, wf \, (1 - u/u_d)^2, \text{ if } u < u_d \tag{3.4}$$

$$D = 0 \text{ (no settling), if } u > u_d \tag{3.5}$$

where A is a parameter defining the consolidation of the mud on the bottom, u_c and u_d are threshold velocities for respectively entrainment and deposition, n is a constant with value between 2 and 4, C is the suspended solid concentration, and w_f is the settling velocity in still water.

Recent field research (Winterwerp and van Kesteren, 2004; Maa and Lemckert, pers. com.) suggests that in the field w_f in Eq. (3.3) may be much higher, possibly by a factor of 10, than that measured in still water. There are at least two explanations for this. Firstly, the sampling procedure to capture flocs in an estuary in order to measure their settling velocity in still water, may mechanically destroy the large flocs. Secondly, the turbulence affects the floc size by processes of aggregation and break-up (see Fig. 3.10 described later), hence the lack of turbulence in still water leads to a different floc size.

In the presence of waves the value of A in Eq. (3.3) is greatly increased by wave-induced pore pressure build-up until the excess pressure force is enough to carry the weight of the sediment overburden; at that time the mud is fluidised and readily eroded (Maa and Mehta, 1987; Wolanski and Spagnol, 2003). Hence, provided there is sufficient time, even small waves can fluidise massive amounts of mud and generate the migration of large mud shoals. By comparison sand banks are much more stable.

Just like for sand, tidal pumping also exists for mud because of the non-linear dependence of the erosion and deposition rates on the water velocity (Eqs. 3.2–3.5). Tidal bores are the extreme case of a tidal asymmetry. However a tidal bore is short-lived and in practice it contributes little to pumping mud landward (for the Humber Estuary, U.K., see Uncles et al., 2006; for the Daly Estuary, Australia, see Wolanski et al., 2006a). However, as was shown in the Daly Estuary, the bore is followed by a 20-min long period of intense macro-turbulence where the water velocity increases by an additional $0.6–1 \, \text{m s}^{-1}$ in 8-sec long events that are due to the passage of eddies (Wolanski et al., 2006a). This doubles the amount of mud eroded from the bottom, and this resuspended mud is transported landward by the flood tide. This mud settles up-river at slack tide and most of it is not resuspended at the following ebb tide; thus the mud has been transported landward. Hence, by generating macro-turbulence in its lee, the tidal bore indirectly contributes to half of the net tidal pumping.

Field studies reveal that Eqs. (3.2–3.5) need to be changed to be applicable to estuaries where the biology is important. Settling is decreased when zooplankton feed on organic matter in the flocs and in so doing break up the large flocs into smaller ones (Fig. 3.9); this decreases the floc size and thus the settling velocity. Algae mat harden the surface of mud banks and protect the sediment from erosion (Fig. 3.10; Andersen et al., 2005; Lumborg et al., 2006). Conversely, animals burrowing in the mud, such as the mudsnail *Hydrobia ulvae*, both destabilise the consolidated mud and make it more erodible, and stabilize the mud by compacting it into fecal pellets. Settling is also enhanced by pelletisation of flocs filtered out of suspension by feeding bivalves (Fig. 3.10). The settling velocity w_f depends on floc size and density of the floc and these depend on the biology. This not just of academic interest; the differences of the settling velocity of suspended mud flocs with/without the biology are a factor of ten (Fig. 3.11a).

For SSC $< 1 \, \text{g} \, \text{l}^{-1}$ the settling velocity increases with increasing SSC values (Fig. 3.11a). For SSC $> 5 \, \text{g} \, \text{l}^{-1}$ the settling velocity decreases with increasing SSC values, in a process called hindered settling. This happens because for sediment to settle an equal volume of water has to move upward. This dewatering happens by the water moving upward in micro-channels through a dense network of settling mud flocs (Fig. 3.11b; Wolanski et al., 1992a). The settling mud flocs align themselves in a train containing typically 6–10 flocs, each floc settling in the lee of the preceding floc. The train floc entrains its interstitial water downward. There is a micro-turbulent shear layer separating the upward moving water in the micro-channel from the downward moving floc train and its interstitial water. If the

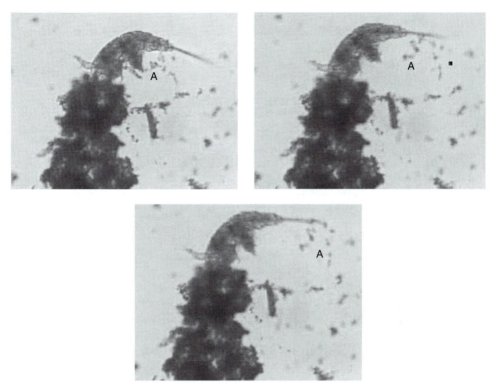

FIGURE 3.9. From left to right, photographs 0.2 sec apart showing a copepod digging out material from a muddy marine snow floc. The dug out material (labelled A) is ejected towards the right.

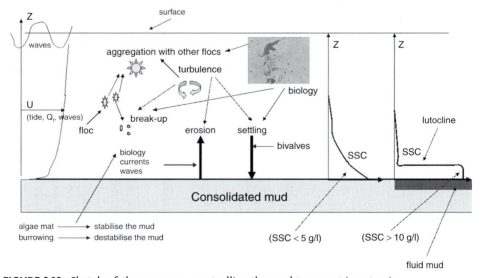

FIGURE 3.10. Sketch of the processes controlling the mud transport in estuaries.

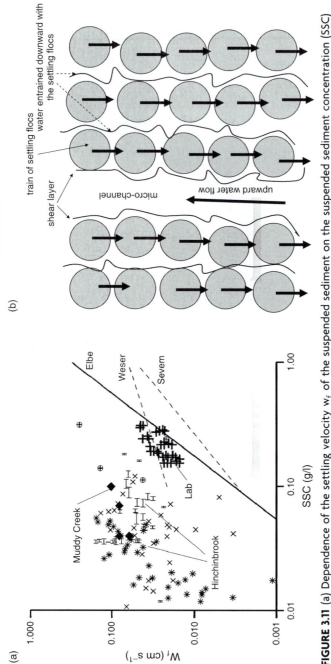

FIGURE 3.11 (a) Dependence of the settling velocity w_f of the suspended sediment on the suspended sediment concentration (SSC) for SSC $< 1\,g\,l^{-1}$. The biology is a dominant process because (1) w_f is larger in tropical estuaries (Hinchinbrook Channel and Muddy Creek, Australia) than in temperate estuaries (Elbe, Severn and Weser estuaries in Europe) as a result of higher temperature, and (2) w_f is larger in the field than in the laboratory (Lab = Hinchinbrook mud used in the laboratory). Modified from Syvitski et al. (2005). (b) Settling of mud flocs for SSC $> 5\,g\,l^{-1}$.

water is turbulent, the micro-channel is often blocked by a floc ejected sideway from the floc train. This slows down the dewatering process and decreases the settling velocity.

The location of the muddy marine snow zone varies from estuary to estuary according to the turbidity. It typically requires SSC values less than $100 \, \text{mg} \, \text{l}^{-1}$ for sufficient light penetration to allow photosynthesis. Thus in the extremely turbid Fly Estuary, Papua New Guinea, and in King Sound, Australia, the muddy marine snow zone is located offshore where the water is clearer to allow photosynthesis, while in the much less turbid Chesapeake Bay, U.S.A., and in Darwin Harbour, Australia, it is located in the upper estuary (Ayukai and Wolanski, 1997; Roman et al., 2001; Wolanski and Spagnol, 2003; Williams et al., 2006; Sanford et al., 2005; Marshall et al., 2006; Wolanski et al., 2006b).

The stability of sediment against erosion and resuspension is even more complex in intertidal areas due to higher biological activity. Through bioturbation, detritivorous invertebrates make the sediment available for erosion. Worm tubs can stabilise the mud. Through filtering the water by fecal pelletisation they can enhance the deposition of suspended matter. Conversely the amphipod *Corophium volutar* reduces sediment cohesion and make it available for erosion by feeding on the benthic diatoms that bind the sediment particles. Birds feeding on this amphipod help preserve the sediment. Crests of bedforms may be more stable than troughs due to biostabilisation by benthic diatoms (Blanchard et al., 2000; Christie et al., 2000).

Flocs change in size periodically with the tides; they increase in size at small values of turbulence-induced shear G and they break up at high values of G (Dyer et al., 2002). Floc size also depends on the biology (Figs. 3.8 and 3.10). This results in a vertical profile of SSC that is smooth, typically decreasing smoothly with elevation from the bottom for SSC $< 5 \, \text{kg} \, \text{m}^{-3}$ ($5 \, \text{g} \, \text{l}^{-1}$). The vertical profile of SSC is often discontinuous for SSC $> 5 \, \text{kg} \, \text{m}^{-3}$ with a sharp discontinuity that is called the lutocline (Fig. 3.10). Because suspended sediment increases the density of the water, the water density changes across the lutocline. This reduces the vertical diffusion coefficient K_z (Fig. 2.6b), and in turn this inhibits turbulent mixing and erosion of the lutocline (Fig. 3.13b discussed later). This helps preserve a fluid mud layer near the bottom.

The pathways for fine sediment in shallow, muddy, mesotidal and macrotidal estuaries are complex, with physics and biology equally important. Flocculation and tidal pumping generate a turbidity maximum zone (Fig. 3.12). Fine sediment is largely retained in the estuary (Uncles et al., 2006). New riverine sediment is diluted by old riverine sediment in a 'mud bath'. This process is particularly important for turbid estuaries in industrialised countries that have cleaned up their point sources of industrial pollution, yet the estuarine sediment remains polluted. The new, non-polluted mud dilutes the 'old' organic pollutants, heavy metals and radionuclides that are absorbed on the 'old' mud of turbid estuaries (Fig. 3.6; Valette-Silver, 1993; Oldfield et al., 1993; Williams et al., 1994; Gueuné and Winnet, 1994; Leggett et al., 1995).

The pathways of fine sediments in microtidal estuaries and in fjords are much simpler because the sediment is not kept in suspension by tidal motions and it

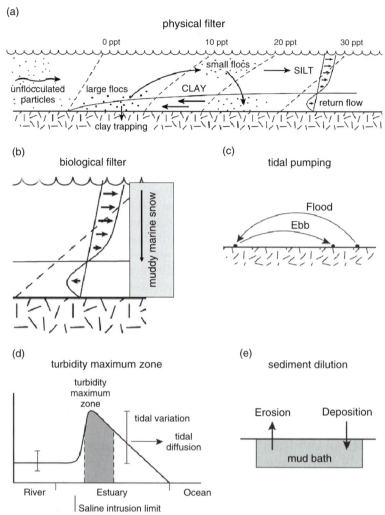

FIGURE 3.12. (a) Riverine mud in suspension generally arrives unflocculated in the estuary. On meeting seawater – salinity is shown in ppt – a physical filter operates. The mud particles flocculate and form the largest flocs near the salinity intrusion limit where the residence time is the longest. The small flocs escape, move seaward, grow larger, settle to the bottom layer and are advected landward by the salinity-induced secondary circulation (the return flow). Silt-based flocs are weaker and readily broken up by turbulence and are preferentially exported. (b) On reaching clearer water in the estuary or in coastal water, the biology becomes important; large, muddy marine snow flocs form and settle down to be brought back landward in the estuary by the salinity-driven secondary circulation. (c) Tidal pumping preferentially drives the sediment landward. (d) These processes result in forming a turbidity maximum zone. Tidal fluctuations of suspended sediment concentration are large. Some sediment is exported seaward by tidal diffusion. (e) The bulk of the fine sediment remains in the estuary, forming a 'mud bath'. 'New' riverine sediment is diluted with 'old' sediment. Modified from Syvitski et al. (2005).

simply settles down. It is continuously capped by new sediment. However, in shallow waters this material can be resuspended by waves.

Fluid mud lubricates the water flow over the bottom by diminishing bottom friction (McAnnally and Hayter, 1990; Mehta and Srinivas, 1993; King and Wolanski, 1996). Thus the apparent bottom friction coefficient in an estuary with fluid mud is measurably smaller than that in an estuary with sand or consolidated mud.

3.2.4. Engineering implications

Water density ρ increases linearly with increasing SSC. In muddy estuaries, water density commonly increases with increasing water depth, often in a series of steps (Fig. 3.13a). There is often no clear bottom in a muddy estuary. The 'classical' 210 kHz echo sounders used to produce bathymetric maps see the bottom at $\rho = 1.06 \, \mathrm{t \, m^{-3}}$. Lead lines and 30 kHz echo sounders see the bottom at $\rho = 1.3 \, \mathrm{t \, m^{-3}}$. Ships can still move through mud up to a density of $1.2 \, \mathrm{t \, m^{-3}}$. For navigation the engineering implication is that it is only necessary to dredge mud with $\rho > 1.2 \, \mathrm{t \, m^{-3}}$.

When the water is very muddy and the currents sluggish, the turbulence is inhibited by the buoyancy due to SSC gradients and incomplete mixing results (Fig. 3.13b). The suspended mud is distributed in a patchy manner throughout the water column (Fig. 3.13c) and mud boils are visible at the surface (Fig. 3.13d).

The 3-dimensional currents in meanders (e.g. Fig. 2.12a) sort the mud from the sand and create mobile shoals. These shoals in turn create three-dimensional currents (e.g. Figs. 2.13 and 2.14) that move the shoals. This complicates navigation because the bathymetry can change quickly and needs frequent surveying, and navigation markers need frequent repositioning.

These internal currents also separate mud from sand because the smallest particles (mud and silt) are uplifted toward the surface in an upwelling while the heavier sand particles remain on the bottom. The estuary bed can become entirely sandy and the sloping banks muddy (Wolanski, 1992).

3.2.5. Biological implications

Mud can be resuspended by wind waves and exported to the estuary and coastal waters as a bottom-tagging nepheloid layer (Fig. 3.14). Mud escaping from mud banks and reclamation areas can thus broadcast over long distances and impact sensitive ecosystems such as seagrass and coral reefs.

This exported mud can also be beneficial by supporting penaeid shrimps in offshore mud banks (Alongi and Robertson, 1995). It can also be beneficial by generating high turbidity and low light levels, resulting in fewer HABs (harmful algae blooms) in highly eutrophicated estuaries such as the Pearl River estuary in China (Zhang et al., 2006). However this service is at the cost of a degraded environment. This mud is harmful to seagrass because it decreases turbidity (Duke and Wolanski, 2001). Indeed, a loss of seagrass is commonly the first indication of coastal waters getting muddier. Muddy marine snow flocs are also

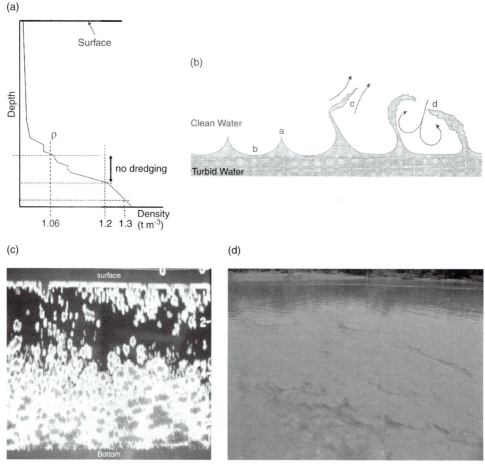

FIGURE 3.13. (a) Typical vertical profile of water density as controlled by SSC in a muddy estuary. (b) Sketch of internal waves deforming the lutocline to produce sharp crests (**a**) and flat troughs (**b**), a single filament entrained upwards by a turbulent eddy in the upper layer (**c**), and filament pair (**d**) entrained upwards by a turbulent jet impinging on the density interface. Adapted from Jiang and Wolanski (1998). (c) Patchiness revealed by a 210 kHz echo sounder stationary for 5 min in the muddy Fly Estuary, Papua New Guinea. On the right, the depth is shown in m. (d) Photograph of a mud boil in the turbid Daly Estuary, Australia.

most harmful to small benthic organisms such as barnacles and coral polyps because they stick on them and literally choke them to death (Fig. 3.15a and b; Fabricius and Wolanski, 2000; Fabricius et al., 2003). Muddy marine snow flocs even stick to live phytoplankton and zooplankton (Fig. 3.15c) and stress them by weighting them down. Muddy marine snow flocs smother and degrade coral reefs in estuaries as well as in coastal waters up to tens of km away from the river mouth (Fabricius and Wolanski, 2000; Wolanski et al., 2003a and 2005; Fabricius et al., 2003).

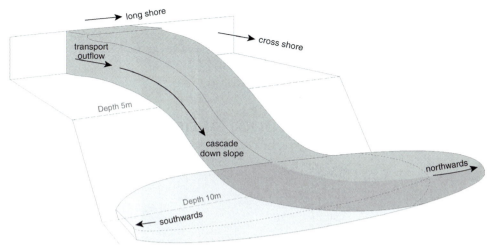

FIGURE 3.14. Formation of a nepheloid layer escaping from a mud bank or a reclamation area. Adapted from Brinkman et al. (2004).

3.3. NET SEDIMENT BUDGETS

3.3.1. The age of estuaries

Land clearing increases soil erosion and thus the riverine sediment delivery to the estuary (see Table 1.1). Large dams trap sediment but also decrease large river floods that previously flushed sediment out of the estuary. If the inter-tidal wetland decreases as a result of land reclamation, the tidal asymmetry is lost, self-scouring stops and the tidal creek silts. In the case of the Klong Ngao Estuary in Thailand where half of the mangrove land was reclaimed by shrimp farmers, the estuary silted within 5–10 years so that it now dries up completely at low tide. In its natural state it was navigable even at low tide (Wattayakorn *et al.*, 1990). Thus the wetland vegetation is essential to maintain navigable channels.

All these human activities accelerate the aging of estuaries by enhancing the trapping of sediment in the estuary, particularly so for a muddy, meso- or macro-tidal estuary (Fig. 3.12). The resulting time scale for aging of an estuary is predicted to be thousands of years for pristine river catchments, hundreds of years for moderately impacted catchments, and tens of years for heavily impacted catchments (Fig. 3.16). This is because these estuaries import sediment from the sea while occasional large river floods flush out sediment seaward (Fig. 3.2) and thus retard senescence. By suppressing large floods dams accelerate the senescence of estuaries; such is the case of the Ord Estuary in Australia and such is the predicted future of the Mekong Delta, Vietnam, following construction of dams in China (Syvitski et al., 2005; Hoa et al., 2007).

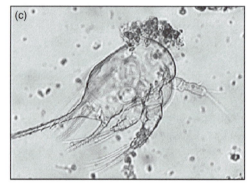

FIGURE 3.15. Photographs of muddy marine snow flocs adhering to (a) the tentacles of a barnacle, (b) a live coral, and (c) a live copepod. In (b) the top of the coral branch contains hundreds of polyps that are completely covered and smothered by the large floc.

3.3.2. Net erosion or progradation

Deforestation and poor land use in the river catchment increase the riverine sediment load and increase seaward coastal progradation. The rapid sedimentation of estuaries and the formation of new muddy river deltas, formed over a few decades, are such examples. The impact can be dramatic, and is sketched in Fig. 3.17a for the case of the Cimanuk River in Java, Indonesia (Wolanski and Spagnol, 2000). Deforestation of the mountainous upper catchment in World War II has resulted in two silt waves, one at the base of the coastal strip and the other one in the river delta; these two silt waves are progressing towards each other and have raised the bed and flood levels by up to 4 m in 40 years

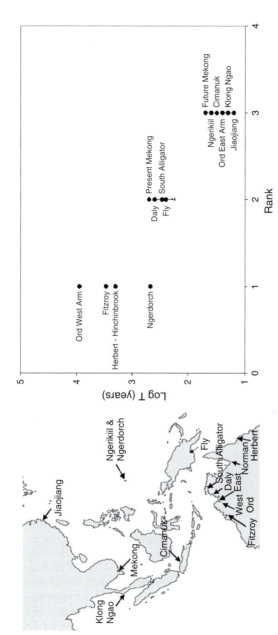

Figure 3.16 Location of macro- and meso-tidal, muddy estuaries in the Asia Pacific region, and the aging time scales for three different levels of human activities in the river catchment. A rank of 3 means that at least 25% of the catchment has been cleared, or that the flow is regulated by at least one large dam, or that the freshwater flood plains become tidally inundated as a result of a sea level rise. A rank of 1 means a largely undisturbed catchment. A rank of 2 means a disturbance level between those for ranks 1 and 3. Modified from Wolanski (2006b).

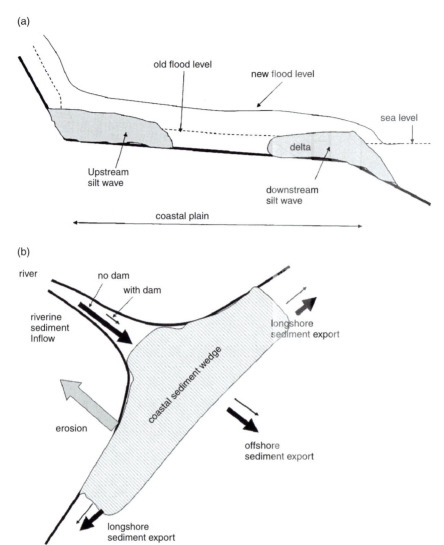

FIGURE 3.17. (a) Sketch of the changes in the coastal plains of the Cimanuk River, Indonesia, following deforestation of the mountains during World War II. In 40 years the river in the coastal plain has silted by up to 4 m vertically, increasing flood levels in the coastal belt also by 4 m, and the muddy delta has grown seaward by 5 km. Modified from Wolanski and Spagnol (2000). (b) The net movement of the coast as seaward progradation or landward erosion is determined by the net budget of the coastal wedge. This budget is the balance between the riverine sediment inflow, and the inflow/outflow of sediment from that wedge as longshore and offshore exports or imports at wave-dominated settings. The riverine sediment inflow itself is the balance between the outflow of sediment landward back into the estuary by tidal pumping and the inflow of sediment in the coastal sediment wedge during river floods. Practically all the riverine sediment is exported from the estuary into the coastal wedge when the estuary has freshwater up to the mouth at least during floods, such as in the Amazon River (Geyer and Kineke, 1995), New England estuaries (FitzGerald et al., 2005), the Mekong Estuary (Wolanski et al., 1996b), and the Daly Estuary (Wolanski et al., 2006a).

since World War II. In the coastal plains the bed level is now higher than the surrounding coastal plains that are heavily populated. To prevent flooding of the coastal plains, 3–4 m high levees have been constructed along the lower 30 km of the river. When a levee breaks, catastrophic floods result.

Another example of rapid, human-induced, estuarine and river changes in geomorphology as a result of land clearing is that of the Thames River floodplains, U.K. Deforestation and ploughing started about 2,600 yr B.P. and created a 0.5–1 m layer of clay alluvium overlying the original soils comprising sandy limestone gravel (Hazelden and Jarvis, 1979).

Large dams can have the opposite effect by trapping much of the sand in the reservoir; this creates coastal erosion by starving the coastal sediment wedge (Fig. 3.17b). It also increases muddiness and turbidity of estuarine waters. Such examples abound (Wolanski et al., 2004a; Syvitski et al., 2005; Kim et al., 2006). For instance about 90% of the Nile River sediment is trapped by the Aswan High Dam; as a result, coastal erosion is intense – the Rosetta and Damietta promontories are eroding at the rates of 106 m year^{-1} and 10 m year^{-1} respectively. The Ribarroja-Mequinenza dam on the Ebro River, Spain, traps about 96% of the riverine sediment; this has led to coastal recession – reversing the previous seaward progradation of the delta. Water diversion from China's Luanhe River has decreased the riverine sediment load by 95% and resulted in its delta's recession at a rate of about 17.4 m year^{-1}. The Mississippi River suspended sediment load has decreased by about 40% between 1963 and 1989, and this may be the major cause for the recession of the Mississippi deltaic coast. The on-going, rapid shoreline retreat in most segments of the Atlantic coast of Portugal is also caused by dams. It is also likely that the Three Gorges Dam in China, under construction, will generate coastal erosion and recession (Wei et al., 2007). The Danube River used to discharge about 6.9×10^5 t of sediment year^{-1}. Since construction of two large dams (Iron Gate I and II) on the river, the riverine sediment flux to the Black Sea has about halved. The Black Sea coast of the Danube Delta is eroding at a rate of 4–7 m year^{-1} as a result of these two dams as well as pollution that destroyed mussel beds in coastal waters, the mussels contributing up to 50% of the beach sediment (Ungureanu and Stanica, 2000).

3.3.3. Formation of mudflats

Non-vegetated mudflats comprise clay and silt mixed with coarser sediments. These mudflats are exposed at low tide and inundated at high tide. They are formed and they accrete when turbid estuarine waters inundate the mudflat on a rising tide and deposit sediment over high water slack, and then do not remove this sediment on the subsequent falling tide (Pethick, 1986). These mudflats have a drainage consisting of a creek and feeder branches, this pattern is prominent in macro- and meso-tidal estuaries (Fig. 3.18a and b) and very subtle in micro-tidal estuaries where it exists nevertheless. The drainage pattern focuses the largest currents in the creek and the smallest current in the surrounding flats where sediment can accumulate. Thus the principles governing water currents and sedimentation in estuaries (see chapters 2 and 3) also apply to mudflats

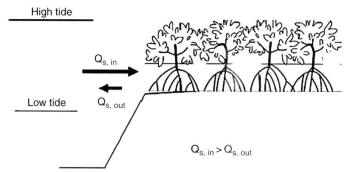

FIGURE 3.20. Tidal wetlands trap fine sediment because the vegetation enhances settling of fine sediment in suspension. Therefore the inflow $Q_{s,in}$ of suspended fine sediment at rising tide is larger that the outflow $Q_{s,out}$ at falling tide. This sketch shows mangroves but the same process applies also for saltmarshes.

mud bank is stable (i.e. is not eroded by storms or tides). Muddy waters enter intertidal areas, deposit some of their suspended sediment in quiet zones in the wake of the vegetation near slack high tide in the wetlands, and return to the estuary with less sediment. The difference between the mud that comes in and goes out is the sediment trapping (Fig. 3.20). A mangrove that covers 3.8% of the river drainage area traps 40% of the riverine mud inflow, the rest contributes to estuarine siltation (20%) and is exported to coastal waters (40%; Victor et al., 2004 and 2006). This relationship is independent of land use in the river catchment because it holds true for developed and undeveloped catchments (Fig. 3.21a). Mangroves fringing muddy open waters are also effective in trapping large amounts of mud from sheltered, coastal waters, up to 1,000 tons km^{-2} yr^{-1} (Wolanski et al., 1998b). The reason for this enhanced sedimentation is the complex flow field around the vegetation that generates zones of flow stagnation at scales of 10 cm (Fig. 3.21b) and 1 cm (Fig. 3.21c) that are dictated by the vegetation, which is where the sediment settles preferentially (Furukawa et al., 1997). This sediment is later evenly distributed by bioturbation.

For non-sheltered coastal waters, mangroves fringing mud-poor reef waters are nutrient-starved and stressed; they are stunted and weakened. Woodborers preferentially attack them because stressed trees generate less tannin than healthy trees (K. Furukawa, pers. com.). The establishment of trees in such intertidal areas is thus very slow.

Similar estimates of the net sediment budgets also exist for saltmarsh-fringed estuaries. Saltmarsh accrete by sediment-laden estuarine waters spreading over the marshes where the sediment drops out of suspension. Sedimentation forms small levees near the banks of the tidal creek, and vertical accretion is achieved by lateral migration of the creek (Bridges and Leeder, 1977; Pethick, 1992; Pethick et al., 1992). Plants play a key role in enhancing the sedimentation. Sedimentation is increased when the saltmarsh vegetation is taller than 8 cm (Boorman et al., 1998). Among the taller species, *Spartina anglica* and *Puccinellia maritima* enhance sedimentation in established European saltmarshes (Thompson, 1990;

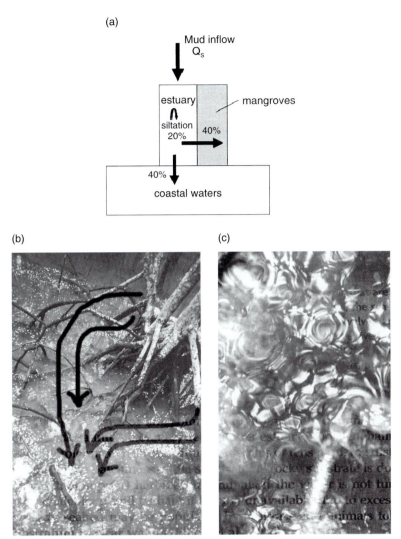

FIGURE 3.21. (a) Fine sediment budget for meso-tidal and macro-tidal estuaries fringed by a mangrove swamp that cover 3.8% of the catchment area. Redrawn from Victor et al. (2004 and 2006). (b) Time-lapse photograph of surface tracers and the inferred tidal currents around a mangrove tree, showing a 10 cm wide stagnation zone in the lee of the trees and roots. (c) Vertical photograph of cm-scale turbulent eddies in flows through mangrove vegetation.

Gray et al., 1991; Sanchez et al., 2001). *Puccinellia maritima* plays a role in enhancing sedimentation and in the establishment of micro-topography in the pioneer zones of lower saltmarshes (Langlois et al., 2003). *Puccinellia sp.* is also involved in the geomorphogenesis of the lower marsh, as evidence by the positive linear relation between the area of hummocks and the abundance of *Puccinellia sp.*, leading to the formation of hummocks on which it is dominant. The sedimentation is very rapid (up to 82 mm yr^{-1}) in the pioneer zone (Scholten and Rozema, 1990;

Langlois et al., 2003). *Puccinellia sp.* is particularly efficient at stabilizing sandy sediment because of the density of its root system and the speed of its spatial spread by rhizome propagation, as well as its tolerance to burial (Richards, 1934; Dijkema, 1997; Langlois et al., 2001). Once the micro-topography has been established, the rates of rise of the substrate and vegetation succession both accelerate. *Spartina anglica* does poorly in sandy sediments, with an abundance typically less than 10% ; waves inhibit the natural spread of *Spartina anglica*, and ergot fungus affects seed viability and germination ability (Groenendijk, 1984; Gray et al., 1990).

Rates of sedimentation recorded at U.K. saltmarshes average 4.3 mm yr^{-1} over two years of observation, with large local variations between 6.2 mm $month^{-1}$ of erosion and 4.1 mm $month^{-1}$ of accretion (Lefeuvre, 1996). Episodic storm events, and not regular tidal inundation, account for 90% of that accretion. A similar result is also found in U.S. saltmarshes (Cahoon et al., 1996).

During spring tides there exists a marked tidal asymmetry of the currents in tidal creeks that drain tidal wetlands, the peak ebb tidal currents at the mouth of the creek being measurably larger than the peak flood tidal currents (Fig. 2.11; Wolanski et al., 1980 and 2001; Kjerfve et al., 1991; Wolanski, 1992). Thus tidal creeks fringed by intertidal wetlands are self-scouring.

Even prograding coasts have periods of recession (erosion). They erode during storms and recover sediment during calmer periods (Pethick, 1992). Over distances of only a few hundreds of m there may be divergent temporal trends in wave exposure and in erosion and sedimentation. The location of eroding and prograding sites thus can vary enormously, sometimes at a distance of only of a few tens of m, within an estuary and along the coast as a function of local exposure of the coast to waves and wind (Davies and Johnson, 2006).

Biological processes may initiate the formation of tidal creeks. Indeed, by digging burrows under saltmarsh vegetation, the burrowing crab *Chasmagnathus granulatus* causes the area to become slightly depressed (Perillo et al., 2005; Minkoff et al., 2006). A depressed ring is formed, which continues growing outwards, joining other rings to start to develop a depression. Tidal currents are funnelled in this depression and they excavate a tidal creek.

3.4. THE SIZE OF THE MOUTH

The cross-section area A_c (in m^2) of the mouth at low tide is determined by the balance between sediment export and sediment import. For sandy, tidally dominated estuaries, this was first quantified empirically by O'Brien (1931) as,

$$A_c \approx 5.74 \times 10^{-5} P^n \qquad (3.6)$$

where P is the tidal prism (i.e. the volume of seawater entering the estuary at rising tide, in m^3) at spring tide, and n is a constant (in the range 0.95–1.23 according to the sediment size). This relationship is the basis for a number of modern empirical formulae relating A_c to P_c (Gao and Collins, 1994). These formulae are helpful for engineering applications, such as determining if an inlet may widen and thus damage housing. They neglect the transient state of estuaries, which includes river floods that can scour and open the mouth of an estuary and of waves at sea that can transport enough sand to intermittently block an estuary. They neglect the transient nature of many estuaries that are commonly in a recovery mode from the last big flood or storm at sea. They are least applicable in muddy estuaries.

CHAPTER 4

Tidal wetlands

4.1. DESCRIPTION

Wetlands have standing water for some period of the year and vegetation that is adapted to or tolerant of saturated soils. Saltwater estuarine wetlands are present, commonly located in the lower reaches (Fig. 1.9). Freshwater estuarine wetlands also occur; these are located in the freshwater part of the estuary and also landward of saltwater wetlands in the saline region of the estuary (Fig. 1.9). Tidal wetlands occur where the rate of accumulation of sediments is equal to or greater than the rate of land subsidence and where there is adequate protection from waves and storms. They are commonly vegetated where the substrate is above mean sea level and tidal inundation is frequent enough to prevent hyper-salinisation. Tidal wetlands include salt marshes, mangroves, and mud flats. Traditionally they have been studied for their vegetation and fauna. Only recently has the focus changed to understanding the wetland as an ecosystem linked with the adjacent estuary (Mitsch and Gosselink, 2000; Wolanski et al., 2001 and 2004a; van der Valk, 2006). Tidal wetlands play a critical role in determining the robustness of the estuary, by selectively trapping fine sediments, influencing the water residence time, sequestering nutrients and pollutants, and converting excess nutrients within the water column into plant biomass. These points are described below.

The world *mangrove* probably comes from the Portuguese word *mangue*, which means "tree", and the English word "grove" for a stand of trees. Mangroves worldwide cover an area of about $240,000\,km^2$ and are the dominant coastal wetland in subtropical and tropical regions where frosts do not occur. There is generally little understory in undisturbed mangroves and extensive fern cover in disturbed mangroves. The topography, drainage pattern and the resulting hydrodynamics, generate three different mangrove swamp types (Fig. 4.1; Lugo and Snedaker, 1974). The largest trees ($\approx 20\,m$) and the highest vegetation density including density of roots and pneumatophores are generally found in R-type (i.e. riverine) mangroves. These are located along sheltered, tidal estuaries with input of freshwater and nutrients from uplands. R-type mangroves are usually drained by tidal creeks (Fig. 4.2a) that themselves may branch out in dentric channels. There is often a zonation of trees' species both along the creek and

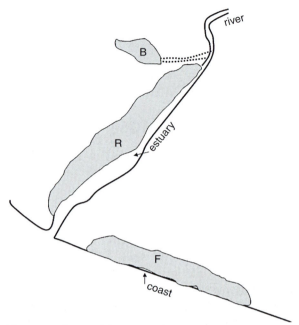

FIGURE 4.1. Typical location of Basin (B), Riverine (R), and Fringe (F)-type mangroves. Dwarf mangroves (not shown) are usually restricted to sandy soils or limestone substrate.

across the wetlands in apparent response to tidal flooding and frequency, soil salinity, soil permeability, and nutrient limitation (Fig. 4.2b).

The trees are smaller ($\approx 13\,\mathrm{m}$) and the vegetation density is lower in F-type (i.e. fringe) mangroves. These are generally located along coastlines that are usually protected by a coral reef, a shoal, or a headland from large sea waves. The trees are even smaller ($\approx 9\,\mathrm{m}$) in B-type (i.e. basin) mangroves. These are located in inland depressions where flushing may be inhibited by the presence of a sill and where water stagnates. There are also (not shown) dwarf mangroves which are isolated wetlands, sometimes only clump of trees, limited to typically $2\,\mathrm{m}$ in height, growing in stressful environments, such as nutrient-poor or hypersaline waters and in sandy soils or in cracks in the limestone substrate.

Saltmarshes (Figs. 4.2 c and d) are distributed worldwide, mainly in areas subject to frosts where mangroves do not grow. Thus they are most common in middle and high latitudes. The vegetation comprises mainly salt-tolerant grasses and rushes. Mud and epiphytic algae are also common. They are also occasionally found in tropical areas on the landside of mangrove swamps. Like R-type mangroves they are also usually drained by tidal creeks (Fig. 4.2c). The shape of saltmarsh-fringed tidal creeks is similar to that of mangrove-fringed tidal creeks, as is apparent from comparing Figs. 4.2 a and c. Like mangroves their vegetation also shows zonation patterns (Fig. 4.2d). For both mangroves and saltmarshes the vegetation is the thinnest in the areas of lowest elevation where accretion occurs and the vegetation is colonizing the area (Fig. 4.2 e and f).

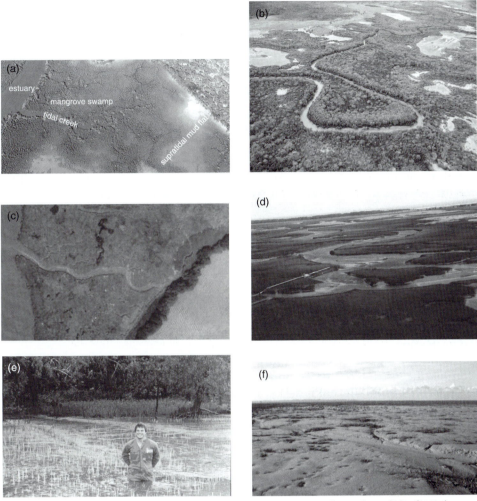

FIGURE 4.2. Aerial photographs of (a) a typical R-type mangrove-fringed creek (Rufiji Delta, Tanzania), (b) vegetation zonation patterns in a R-type mangrove (Mary Estuary, Australia), (c) a typical saltmarsh-fringed tidal creek (Tamar Estuary, U.K.), and (d) a saltmarsh (North Inlet, North Carolina, U.S.A.). Photographs of mud accretion zones at the estuarine side of tidal wetlands being colonized by (e) mangroves and (f) saltmarsh grass (Fly Estuary, Papua New Guinea, and Mont Saint Michel, France). Photos (a), (b), (c), (d) and (f) were kindly provided by I. Bryceson, D. Williams, R. Uncles, B. Kjerfve, and E. Langlois-Saliou.

Intertidal mud flats are of two types. Mud flats located below mean sea level are wetted daily by the tides and the top layer of the mud is commonly soft and unconsolidated (Fig. 4.3a). Mud flats in supra-tidal areas are tidally inundated only at the highest astronomical tides. The mud dries and hardens in-between inundations. A mild slope forms a featureless drainage pattern toward the tidal creek whose fringes are vegetated only in the lower reaches; the tidal creek drains to the estuary (Fig. 4.3b).

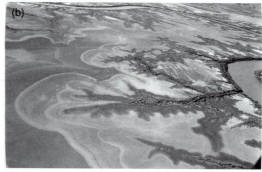

FIGURE 4.3. Photographs of (a) a mud flat located below mean sea level, and (b) a supra-tidal mud flat. The vegetation cover is minimal. (Keep and Ord estuaries, Australia).

4.2. HYDRODYNAMICS

The peak velocities in wetland-fringed tidal creeks can be large ($\approx 1\,\mathrm{m\,s^{-1}}$) while only 10 m away in the vegetated wetland the maximum velocity is small ($\leq 0.08\,\mathrm{m\,s^{-1}}$; Wolanski et al., 1980; Kjerfve et al., 1991; Mazda et al., 2005). This is in the tidal creek fully non-linear, open water hydrodynamics prevail, while in the vegetated wetlands the currents are reduced by friction of the flow around the vegetation.

To model the water circulation in tidal wetlands, the complex bathymetry requires the use of cell-based models where the cells are curvilinear and of irregular shape to fit the bathymetry. The model differentiates open-channel flows in the tidal creek from friction-dominated flows in the vegetated wetland. These models are successful in reproducing the hydrodynamics of a tidal creek fringed by a mangrove swamp or a saltmarsh (Wolanski et al., 1980; Kjerfve et al., 1991). The flow is complex in time and space because the water spreads over a wide area at high tide and it is restricted to a narrow channel at low tide (Fig. 4.4). The tides and the water currents are tidally asymmetric. At rising tide, the water surface along the tidal creek slopes downward from the mouth; therefore the wetlands in the lower reaches are first flooded while those in the upper reaches remain exposed longer. At falling tide, the water surface slopes downward toward the mouth; thus the wetlands in the lower reaches are exposed first while those in the upper reaches remain submerged longer. The water surface slope and the currents in the tidal creek at falling tide are larger than those at rising tide. A similar tidal asymmetry prevails in salt marshes (Fig. 2.11).

Of importance for wetlands is that the velocity vectors in the wetlands are also tidally asymmetric. Indeed, at flood tides the velocities within the wetlands are small, usually less than $0.04\,\mathrm{m\,s^{-1}}$, and directed perpendicular to the banks of the creek. At falling tide the peak velocities are typically twice as large and are oriented downstream intersecting the banks of the creek at an angle of about 30°. This tidal asymmetry is important for the transport of plant litter (e.g. plant detritus for saltmarsh and leaves for mangroves). Indeed at flood tides the floating plant litter is often blocked by the vegetation while at ebb tides the currents

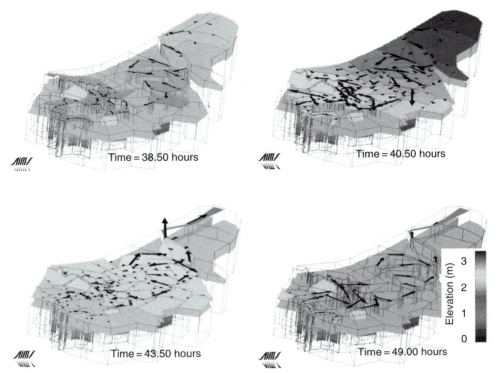

Time = 38.50 hours

Time = 40.50 hours

Time = 43.50 hours

Time = 49.00 hours

Elevation (m)

3

2

1

0

FIGURE 4.4. Snapshots of the predicted water level and the tidal currents in Muddy Creek, a 5 km long meso-tidal, mangrove-fringed tidal creek in Australia. There are two tides per day. The model is based on cells that are fitted to the bathymetry.

can be large enough to force the floating plant litter through the vegetation and export it to the tidal creek (Wolanski et al., 1992b).

In the cell-based model, the water surface slope in the wetlands S_f is friction-induced, thus

$$S_f = n^2 \, Q|Q|/A^2 \, R^{4/3} \tag{4.1}$$

where n is the Manning friction parameter, Q is the flow rate between two cells, A is the cross-sectional area of the flow between cells (i.e. the width times the depth taking into account the area occupied by the vegetation), and R is the hydraulic radius (i.e. in practice the depth). From the model and the field data it is then possible to quantify the degree to which the vegetation contributes to friction. For saltmarshes and mangroves n ≈ 0.1–0.2, which is 4–8 times higher than the value in the creek (n ≈ 0.025; Wolanski *et al.*, 1980; Burke and Stolzenbach, 1983; Kjerfve et al., 1991; Mazda et al., 1997a; Wolanski, 1992). Thus the flow through the wetland is strongly controlled by the vegetation.

The residence time can be estimated by the time it takes for the concentration C of a water-born waste, discharged in the upper reaches of the creek, to reach a steady state distribution in the creek. For a 5 km long mangrove-fringed tidal creek, the residence time is about 7 days (Fig. 4.5).

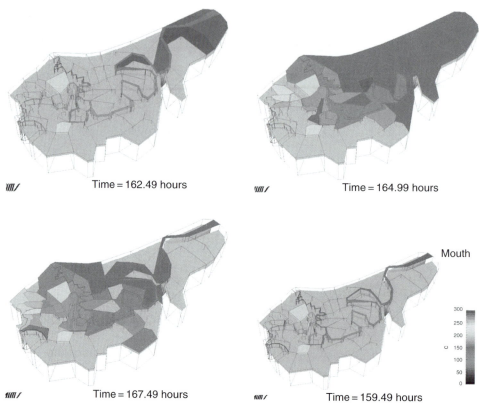

Time = 162.49 hours

Time = 164.99 hours

Mouth

Time = 167.49 hours

Time = 159.49 hours

FIGURE 4.5. For Muddy Creek at the comparable times in the tide cycle as in Fig. 4.4, snapshots of the distribution of the concentration C of waste discharged in the upper reaches of the creek.

4.3. WAVE ATTENUATION BY WETLAND VEGETATION

Mangroves absorb wave energy as a result of wave-induced reversing and unsteady flows around the vegetation. They can thus protect the coast from wave erosion by absorbing wave energy through the drag and inertial forces (Massel et al., 1999). Probably the best data set on this process is that of Mazda et al. (1997b and 2006) at the muddy coast of Vietnam where *Kandelia candel* and *Sonneratia* mangrove plantations have been created over a wide intertidal shoal offshore as a coastal defence against typhoon waves. The plantation is 1.5 km wide (toward offshore) and 3 km long (along the coast). The wave reduction r per 100 m was 17–60% (Table 4.1). The wave height of 1 m at the open sea was reduced to 0.05 m at the coast, enabling aquaculture ponds behind a 2 m high coastal levee (Fig. 4.6a). Without the sheltering effect of mangroves the waves would arrive at the coast with wave height of 0.75 m and the levee would have been eroded and breached (Fig. 4.6b). Because of their pneumatophores, the rate of wave reduction is much higher by up to a factor 3 for *Sonneratia* forests than for *Kandelia candel* forests. A typhoon created a storm surge that flooded the levee. The levee

TABLE 4.1. Wave reduction (r, in %) per 100 m of adult mangrove plantation (Data from Mazda et al., 1997b and 2006).

Mangrove species	Water depth (m)			
	0.2	0.4	0.6	0.8
Kandelia candel	20	20	18	17
Sonneratia	60	40	30	15–40

The value of r without mangroves was about 5% next to the *Kandelia candel* site and 10% next to the *Sonneratia* site.

(a)

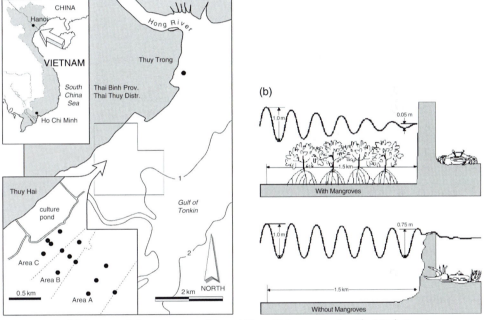

FIGURE 4.6. (a) A map of the mangrove-fringed Thuy Hai coast in the Thai Binh Province, Vietnam. Groups A, B and C are mangrove plantations comprising respectively 0.5 year-old trees, 2–3 year-old trees and 5–6 year-old trees. The symbols • indicate the field measurements sites of tides, waves and currents. (b) A sketch of the wave field at that site (top) with and (bottom) without mangroves. Adapted from Mazda et al. (1997b).

survived and protected the coast because there were negligible swell/waves in the shadow of the mangroves. This is only possible when the forest itself is protected by a wide intertidal shoal that reduces the typhoon waves to 1 m or less; otherwise the trees themselves would be uprooted by the waves.

In temperate countries saltmarsh vegetation plays the role of mangroves in the tropics by absorbing wave energy (Moller and Spencer, 2002). In the absence of trees, the waves that can be dissipated by the grass vegetation of saltmarshes are much smaller than for mangroves. These are locally-generated, small wind waves

FIGURE 4.7. Photographs of the erosion cliff resulting from bank erosion from small wind-driven waves undercutting (a) a salt marsh (Samborombon Bay, Argentina) and (b) a mangrove swamp (Pentecost Estuary, Australia). (c) Speeding, small boat wake eroding the banks of the Danube River delta, Romania. (a) is courtesy of R. Lara.

for saltmarshes, as opposed to typhoon waves for mangroves. The waves can be reduced by 50% in 20 m for *Spartina* plants longer than 25 cm and for a wave height less than 55% of the water depth (Chen et al., 2006). For both mangroves and saltmarshes the intensity of wave attack is very dependent on the coastline configuration (Davies and Johnson, 2006).

Both mangroves and salt marshes are able to absorb wave energy when they are submerged, and thus they protect the coast. However, when they are exposed at low tide they are most vulnerable to erosion by small, wind-driven waves. At that time the waves undercut the banks and undermine the plants in the saltmarsh (Fig. 4.7a) and the trees in the mangroves (Fig. 4.7b); these then topple in the water. An erosion cliff is created. The banks are also susceptible to erosion from speeding, small boat wakes (Fig. 4.7c).

4.4. ECOLOGICAL PROCESSES WITHIN A TIDAL WETLAND

Saltmarshes and mangroves convert excess nutrients into plant biomass that supports an ecosystem. These tidal wetlands store large quantities of nutrients

(Turner, 1993; Valiela and Cole, 2002). In contrast to the primary productivity of eutrophic European estuaries waters where nitrogen (N) is limiting, phosphorous (P) may be the limiting factor in the productivity of European saltmarshes (Doering et al., 1995; Lefeuvre, 1996). There is a large scale removal of P from solution during algal blooms and at such times other elements, such as silicon, may be limiting (Mitsch and Gosselink, 2000; Chicharo et al., 2006).

When the nutrients are brought in from the estuary, this is known as an inwelling. Some nutrients, particularly organic detritus such as plant litter, are exported from the wetlands to the estuary; this is known as an outwelling (Adam, 1990; Lefeuvre and Dame, 1994). There are also other links between the terrestrial and the marine ecosystems. For example, when tidal water flows over a saltmarsh there is a marked increase in the bacterial component of the plankton; mussel beds remove 1/3 of this enhanced production (Newell and Krambeck, 1995). There can also be indirect exports of organic matter in the form of fish and other organisms that come in from the sea to feed within the saltmarsh and then return to the sea.

There are thus many similarities between physical and biological processes in mangroves and saltmarshes. These include trapping fine sediment and pollutants, converting nutrients to plant biomass, enhancing the productivity of estuaries, and serving as a habitat for fish and crustaceans (Fig. 4.8 described below). As shown in section 4.4.3 (see below), supratidal mud flats also enhance estuarine fisheries.

Tidal freshwater wetlands act similarly as saltmarshes, with increased diversity due lesser salt stress and higher internal cycling of nutrients (Mitsch and Gooselink, 2000). Mature tidal freshwater wetlands have large peat reserves, from the decay of plants. Newly emergent marshes have lower sediment organic content and more dependence on flooding water to supply their nutrient needs. Like mangroves and saltmarshes, they are also an important refuge for aquatic fauna including juvenile fish.

4.4.1. Mangroves

In mangroves, forest litter (predominantly tree leaves) is produced at a rate that is the largest for R-type and F-type mangroves, and the smallest for B-type mangroves. Crabs recycle nutrients by carrying a significant amount (up to 50%) of the leaf litter back into burrows, as well as causing an active turnover of the mud (Schories et al., 2003). The hydrodynamics control the fate of the remaining organic detritus (Fig. 4.8a). In R-type mangroves, 94% of the carbon equivalent of that material may be exported to the estuary. In F-type mangroves 42% may be exported and only 21% in B-type mangroves. The rest is lost through decomposition and peat production. This plant detritus exported to the estuary is decomposed in the estuary by fungi, bacteria and protozoa. Combined with primary productivity in the estuary, it supports primary and secondary consumers in the estuary (Fig. 4.8b).

How a mangrove ecosystem functions is still little known quantitatively. Empirical rules abound, such as "*Avicennia* prefers high places", but the scientific

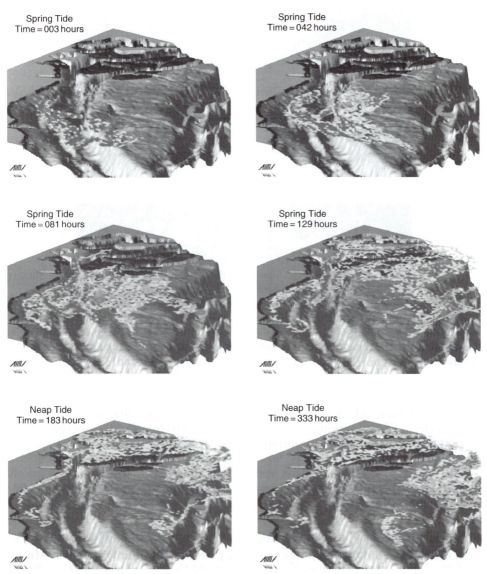

FIGURE 4.11. 3-D bathymetry of the 40 km long Klang Strait, Malaysia, showing the location of channels, shoals, and mangrove swamp, and the predicted distribution of shrimp larvae at different times after spawning over seagrass beds in coastal waters. Modified from Chong et al. (2005).

for them to eat as they are detritus/microinvertebrate feeders (Alongi and Robertson, 1995); they spawn over seagrass beds in coastal waters. After spawning the developing larvae are advected by the water currents. Typically about half of the larvae drift to sea and will die and the other half end up in the mangroves that they use as a nursery (Fig. 4.11; Chong et al., 1996 and 2005). The juvenile shrimps reside in the mangrove creeks at low tide and in the mangroves at high tide where they hide from predators. Only later will they migrate offshore.

4.6. GROUNDWATER FLOW

Groundwater flow rates are commonly much smaller than surface flows. Nevertheless they cannot be neglected because they determine soil properties, especially salinity, and thus the vegetation.

4.6.1. Mangroves

In mangroves, rainfall generally infiltrates directly in the ground and does not create a surface outflow (Wolanski and Gardiner, 1981). Similarly the first surface water observed at rising tides in *Rhizophora* mangroves where the soil has numerous crab holes emerges from the ground through crab holes (Wolanski, 1992). This indicates the importance of groundwater flow.

Groundwater flow drains the water at ebb tide when the water level in the creek is lower than the water table in the substrate (Fig. 4.12a). The groundwater velocity u is calculated using Darcy's law,

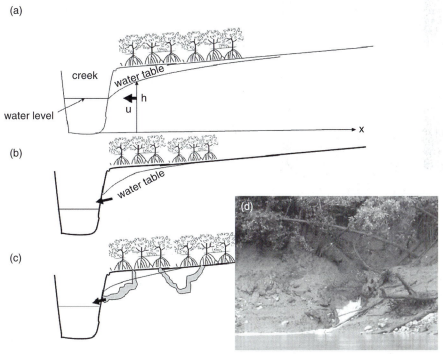

FIGURE 4.12. Sketch of the groundwater flow under a mangrove forest at low tide for (a) continuous and (b) discontinuous flows at the creek's banks. (c) Groundwater flow is accelerated by cavities dug by crabs or left by decaying vegetation. (d) These cavities generate saltwater springs in the creek's banks, such as in this photograph from Darwin Harbour, Australia.

$$u = K_B \, dh/dx \qquad\qquad (4.2)$$

where K_B is the hydraulic conductivity, x is the distance from the creek, and h is the elevation of the water table. In low tide range, the groundwater table remains the same as the water level in the creek even at low tide (Fig. 4.12a; Mazda and Ikeda, 2006). In high tide range, it remains higher; a disconnect results and the bulk of the groundwater outflow occurs on the banks at an elevation located between the water level in the creek and of the groundwater (Fig. 4.12b; Gardner, 2005). For that case, the groundwater outflow to the creek can be calculated by replacing Eq. (4.2) by Laplace's equation for the total head (pressure head plus elevation head; Liggett and Liu, 1983).

A number of engineering formulae exist for calculating K_B as a function of the soil porosity, the sediment particle size and the water viscosity (Mazda and Ikeda, 2006). Values of K_B vary enormously, in the range 10^{-7} to 5×10^{-5} m s^{-1}. The higher K_B values are due to the presence of crab holes and decaying vegetation, producing conducts for groundwater flow (Fig. 4.12 c and d; Susilo and Ridd, 2005; Susilo et al., 2005). The resulting outflow forms a saltwater spring on the creek's banks (Fig. 4.12d).

Groundwater salinity under a mangrove forest seldom exceeds 50, compared to 35–37 in the tidal creek (Fig. 4.9), while the groundwater salinity under a salt pan commonly exceeds 100 and can reach 200 (Ridd and Sam, 1996; Sam and Ridd, 1998; Heron and Ridd, 2003). The reason that salinity is limited in mangrove soils in the dry season is that salt is evacuated by linked physical-biological processes. Some tree species such as *Rhizophora* extract freshwater from saline groundwater and leave the salt behind. This salt is flushed away by groundwater flow that is enhanced by the presence of crab burrows and decaying vegetation (Fig. 4.12 c and d). Other mangrove species such as *Avicennia* extract saline groundwater and evacuate the salt as crystals on their leaves (Duke, 2006). No such mechanisms

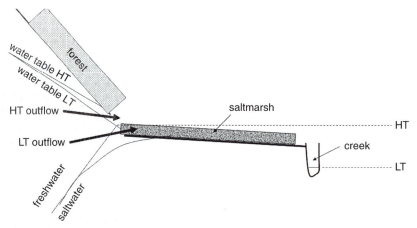

FIGURE 4.13. Sketch of the changes in location of the water table in the uplands, the interface between fresh and saline groundwater, and the freshwater outflow in a saltmarsh, at high tide (HT) and low tide (LT). Interpreted from the data of Gardner et al. (2002).

exist in saltflats because these have no crabs and no trees; thus hypersalinity results from evaporation.

4.6.2. Saltmarshes

Groundwater flow in saltmarshes is quite swift, enhanced by bioturbation and decaying vegetation (Gardner et al., 2002; Gardner, 2005). It results in significant tidal fluctuations of the groundwater salinity interface. It also results in fluctuations of the groundwater table under the saltmarshes as well as upstream because of backwater effects (Fig. 4.13). This controls the location of the transition point between the forests and the saltmarsh.

4.7. PHYSICS-BIOLOGY LINKS

There are thus intricate links between the physics and biology of mangrove-fringed tidal creeks, summarised in Fig. 4.14. Similar links exist for saltmarsh-fringed tidal creeks (Minkoff et al., 2006). The creeks are long and narrow, at times with branching channels. The swamps are wide and heavily vegetated. There are strong tidal flows in the creek as a result of the large amount of water draining in and out of the surrounding swamp. There is tidal asymmetry, resulting in the export of sediment from the creek and thus self-scouring of the creek. Estuarine fine sediment settles readily in the swamp. Because the residence time is large (often > 10 days) in the upper reaches of a tidal creek, water quality is often the worst in those reaches. The complex topography leads to secondary,

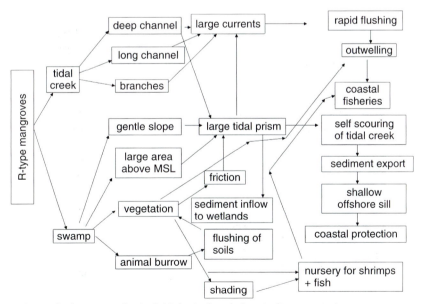

FIGURE 4.14. Links between physical, biological and chemical processes in mangrove swamps. Similar feedback processes also occur in saltmarshes. Modified from Wolanski et al. (1992b).

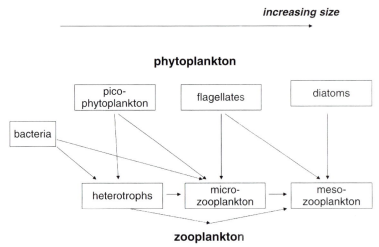

FIGURE 5.2. Predator-prey relationship for phytoplankton and zooplankton. Modified from Allen et al. (2001).

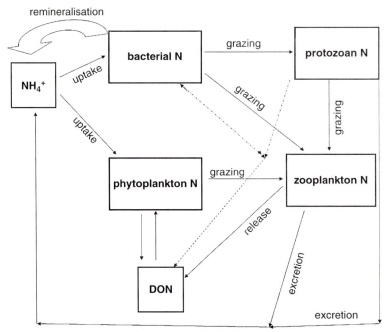

FIGURE 5.3. Sketch of a simple pelagic food web in a turbid, detritus-rich, estuary highlighting the role of the microbial loop in processing nutrients. The first trophic level is the detritus consumers. As in Fig. 5.1., the complete food chain also includes planktivorous fish and bivalves that consume much of the primary production. Many of these consumers feed on particles in suspension or on the bottom. Modified from Alongi (2002).

food web must be added the link with the benthic food web and the link to herbivorous fish. These fish transfer energy and matter from estuarine plants to upper trophic levels through the detrital system. In the microbial loop nitrogen is cycled between phytoplankton, bacteria, protozoa, and zooplankton. The bacterial productivity varies enormously ($0.5–803\,g\,C\,m^{-2}\,yr^{-1}$) as does bacterial biomass ($2–760\,mg\,C\,m^{-3}$; Ducklow and Shiah,1993). In turbid estuaries bacterial production commonly exceeds phytoplankton primary production. Bacteria cannot be considered in isolation from the plankton as there are feedback mechanisms (Fig. 5.3). In addition, viruses can cause mass mortalities of bacteria (Fuhrman and Noble, 1995).

Protozoans remove phytoplankton and they are grazed on by zooplankton (Sanders and Wickham, 1993; Alongi, 2002). Zooplankton feed not only on phytoplankton, protozoa and bacteria, they also feed on detritus particles and associated microbes on organic-rich sediment particles both in suspension and on the bottom (Fig. 3.9). This organic matter contains the remnants of the food web (including dead phytoplankton, macrophytes, and algae), riverine and oceanic detritus, and, by comparison, few living organisms (Tyson, 1995).

To the water-born food web sketched in Fig. 5.3 must be added the contribution of meiofauna and filter feeders such as bivalves (e.g. mussels and clams). These feed on plankton and accelerate the sedimentation of suspended particles by pelletisation, thus also removing their particulate nutrients from the food chain in the water column (Fig. 3.10; Alber and Valiela, 1995; Asmus and Asmus, 1991; Tenore, 1989). Therefore bivalves can reduce phytoplankton densities and control algae blooms. This organic detritus, plus whatever organic material falls from the water column, is not all lost (Fig. 5.4). On settling, the nutrients in the organic

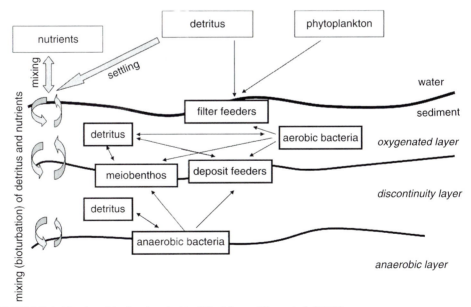

FIGURE 5.4. The benthic food web. Modified from Allen et al. (2001).

macroalgae. The rest (90%) is processed through the detrital system (Alongi, 2002). Several coastal species of fish and crustaceans use the estuary as a nursery ground (Fig. 4.8 b and c; Gillanders and Kingsford, 2002). They affect the uptake and release of organic materials from excretion products and sediment pelleting/burrowing.

Estuarine fish are stressed from changes in salinity but they adapt to it by utilising osmotic regulation, a process that consumes energy. Thus brackish water organisms are usually smaller than their marine relatives.

Estuarine river plumes also support planktivorous pelagic fish species like anchovy (*Engraulis encrasicolus*) or sardine (*Sardinapilchardus*) that yield commercially important fishery catches in coastal waters (Fig. 1.9; Ray 1996; Sklar and Browder 1998; Chicharo et al. 2002).

5.5. ESTUARINE ECOLOGY

As described above, estuaries have multiple sources of nutrients and the loss of nutrients is minimized by a pelagic food web that uses all resources including nutrients, light and space, and minimizes wastage; the loss of nutrients is also minimized by a swift recycling of nutrients between sub-systems. Due to the high residence time, the organic detritus stays long enough in the estuary for its nutrients to become part of the food chain. The water currents minimise stagnation, although they can also increase turbidity and diminish light. Because of high habitat diversity, structural complexity, and varying turbidity, resources are partitioned. This provides ecological stability. Shallow-water habitats in estuaries, including tidal wetlands and seagrass beds, offer food and shelter from predators to juvenile fish (Blaber, 1997; Loneragan et al. 1986; Hannan and Williams 1998; Parish 1989). Varying salinities provide physiological and physical attraction to brackish water and/or to wetland habitats, to fish and shrimp (see chapter 4). In some estuaries where the sediment is stable and the water clear, there are also submerged communities of seagrass and these contribute significantly to the overall productivity (Heip et al., 1995).

For all these reasons, sketched in Fig. 5.5, though not all of them may apply to a given estuary, estuaries are amongst the most highly productive ecosystems on earth (Day et al., 1989; Dollar et al., 1991; Alongi, 2002; Mallin and Paerl, 1992; McLusky and Elliott, 2004). This productivity is not a universal constant; indeed it varies between estuaries from 7 to $560\,\mathrm{g\,C\,m^{-2}\,yr^{-1}}$ (Nienhuis, 1993). It also varies temporally. In temperate estuaries, production peaks during spring and summer in response to solar radiation, warmer waters and possibly higher nutrient availability. In tropical estuaries peak productivity occurs in the post-monsoon season when riverine nutrients are still abundant and water clarity and light availability have improved after the high turbidity during high river flows in the monsoon season (Alongi, 2002).

Productivity varies spatially at a scale of m to km within an estuary because the ecological niches are not uniformly distributed along an estuary (McLusky

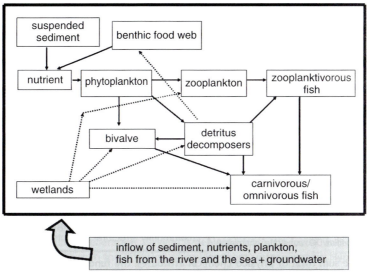

FIGURE 5.5. The pelagic food webs in a typical estuary is driven by nutrients that are provided by suspended sediment, benthic food webs as well as imports from the river, the sea and groundwater. This nutrient is cycled between phytoplankton, zooplankton, bivalves, and fish. Additional fish and plankton is provided by the sea and the river. Much of the riverine plankton becomes detritus when mixed in saltwater. Dying organisms become detritus that is recycled through the microbial loop. By acting as nursery grounds and/or refuges, wetlands (mainly saltmarshes and mangroves) provide detritus, bivalves, and fish to the estuary.

and Elliott, 2004). At scales of km, communities are usually dominated by euryhaline forms (wide salinity tolerance) in the upper estuary, and stenohaline forms (narrow salinity tolerance) near the mouth. At scales of tens of m, the distribution of species reflects the patchiness of the substrate. A rocky substrate is dominated by macroalgae that support herbivorous animals if the water is not turbid and thus light is available; it will be bare if light is unavailable due to excessive turbidity. A sandy seabed may be populated by seagrass and animals tolerant of mobile substratum in clear waters (Elliott et al., 2006).

The ecological niche is the fundamental area that can be occupied successfully by an organism (Fig. 5.6; Elliott et al., 2006). In an estuary, the ecological niche can be very small, sometimes a few m wide, due to the variation of the substrate from sandy to muddy and to rapidly varying turbidities (see section 2.4). Niche breadth indicates the range of environmental conditions suitable for occupation by a species. Some fish species that moves over the seabed are tolerant to many substrate types and thus have a broad niche breadth. On the other hand worms have a narrow niche breadth since they occupy only a special substratum. Similarly, specialised feeders have a narrow niche and non-specialised feeders with a wide range of prey have a wide niche breadth (Naeem, 2002; Elliott et al., 2006). Competition between species or within a species occurs when niches overlap. As sketched in Fig. 5.6, the community ecological niche, i.e. the functioning of a community as an ecological engine, is shaped by the fundamental ecological niche,

upward in daytime. This results usually in a crisper vertical stratification of dinoflagellate blooms than of diatom blooms (J. Lee, pers. com.). About 90% of the HAB species are flagellates (notably dinoflagellates; Smayda, 1997). While nutrients and turbulence can dictate whether either dinoflagellates or diatoms are the dominant species of a particular HAB, generally the most common form of HABs is that due to the single-celled cyanobacterium *Microcystis aeruginosa* in eutrophic, warm, low-salinity waters, especially so if the cells are aggregated by winds, tides, and buoyancy effects (Margalef, 1978; Yanagi et al., 1995; Roelke and Buyukates, 2001; Yamamoto et al., 2002; Robson and Hamilton, 2003). Because the ecology of HABs remains poorly understood, predictive models are still more qualitative than quantitative. Thus operational HABs models use satellites to detect of early signs of HAB and oceanographic models to forecast where such blooms may go (Stumpf et al., 2003; Tomlinson et al., 2004).

The simplest case of estuarine HABs may be that resulting from large dams that totally regulate the river discharge of water and nutrients. Large river dams retain Si and cause a change in the N:P:Si ratios in the estuary. If high nutrient levels and a long residence time prevail in such estuaries, the diatoms first bloom and exhaust the silicon supply while N and P remain abundant in the estuary. Dinoflagellates or cyanobacteria (blue-green algae) do not require Si and can then bloom (Fig. 5.8). Some of these produce toxins. This totally alters the estuarine food web. One reason that such HABs are resilient is that the zooplankton may reject these dinoflagellates as food, possibly due to deterrent chemical compounds, and thus they exert little top-down ecological control (Liu and Wang, 2002).

Some harmful algae have resting stages on the sediment for diatoms (spores), dinoflagellates (cysts) and cyanobacteria (akinetes). These algae have a competitive advantage over populations that cannot survive in poor conditions, as the resting algae can wait for opportune conditions to bloom. A trigger is necessary

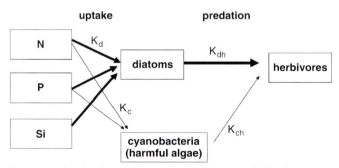

FIGURE 5.8. High riverine loads of N and P into an estuary initially lead to eutrophication and a bloom of phytoplankton (diatom in this figure) because the rate of uptake of nutrients by diatoms (K_d) is much larger than that by cyanobacteria (K_c). In turn herbivores prey on diatoms at a rate K_{dh} larger than the rate that they prey on cyanobacteria (K_{ch}). When the diatoms exhaust the Si, cyanobacteria bloom if N and P remain abundant. Modified from Chicharo et al. (2006).

to generate a HAB. This trigger varies from site to site and may be nutrients from an upwelling event or warm water events, as is observed in the Atlantic coast of France, Spain and Portugal, Chesapeake Bay, the Benguela region and Croatia's Kastela Bay. Similar triggers have been reported for cyanobacteria in the Gulf of Finland. In practice, however, this trigger is only one of a number of meteorological, oceanographic and biological processes necessary to lead to the formation of HABs (Fig. 5.9). If all these other processes do not also occur, the HAB may not develop. These processes are still poorly understood; this makes predicting HABs an art more than a science. These processes may be firstly strong winds that stir the bottom and liberate the cyst from the substrate, secondly a suitable wind direction to generate upwelling to bring nutrients to the surface, thirdly a long time of residence and the absence of predators in surface waters to enable the growth of the harmful algae (Ikeda and Nakata, 2004). Die-off occurs as a result of biological dissipation when the resources are exhausted, and of oceanographic dispersion. During that time, new dormant cysts are deposited on the substrate and these lay dormant waiting for suitable oceanographic and biological conditions to develop the next HAB.

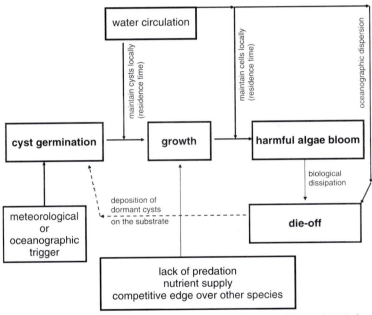

FIGURE 5.9. The cycle of formation and die-off of a harmful algae bloom (HAB) depends on cyst germination triggered by a meteorological or oceanographic trigger, its growth made possible by a long residence time, lack of predation, a suitable nutrient supply, and a competitive edge over other plankton species such as that described in Fig. 5.8. The HAB dies off as a result of biological dissipation when resources are exhausted and as a result of oceanographic dispersion. The cells lay new cysts on the bottom that lay dormant until suitable conditions occur again for cyst germination. Modified from Ikeda and Nakata (2004).

5.7. SEAGRASS AND CORAL REEFS

Seagrass and coral reefs are often found near the mouth of less turbid estuaries and in their coastal waters. Their health is determined by the rate at which fine sediments and nutrients are sequestrated in the estuary, primarily in the turbidity maximum zone and in the tidal wetlands, and to the efficiency of the bacterial loop in the estuary to process the nutrients. The remaining mud and nutrients are flushed out to sea. If human activities increase the sediment and nutrient load, the seagrass beds and the coral reefs degrade. An immediate reason is light attenuation by increased turbidity, as is evident for seagrass beds that progressively die off in waters made more turbid by human activities (e.g. Schoellhamer 1996, and Onuf, 1994, for the case of dredging-induced turbidity; e.g. Duke and Wolanski, 2001, for the case of farming-induced turbidity in coastal waters).

Coral reefs are usually located further offshore than seagrass beds. Seagrasses have roots and are established in softer substrate. They can help protect the reefs by trapping mud and excess nutrients, provided the quantities are not excessive (Fig. 5.10a; Kitheka, 1997; Wolanski et al., 2001).

By contrast, seaweeds have no roots and attach themselves unto solid substrates, such as coral reefs. Any mud that they trap ultimately lands on the reef and kills the coral, and does not protect it (Fig. 5.10b; Fabricius, 2005).

The fraction of a reef substrate covered by live coral naturally fluctuates because some corals are occasionally killed by natural events such as river floods and storms or tropical cyclones. The storms or tropical cyclones kill the corals by mechanic forces (Done, 1992a). The river plumes kill corals by bringing freshwater, fine sediment and excess nutrients to the reefs (McCook 1999; McCook et al., 2001; Wolanski et al., 2003a and 2004b). The empty space is readily colonized by fast growing algae (Fig. 5.11a). The corals can recolonise that space after several years provided that water quality and the quality of the substrate remains good in the recovery period. The length of the recovery period depends

FIGURE 5.10. (a) The trapping of sediment by seagrass growing in soft substrates between the estuary and the coral reef helps shield coral reefs from riverine sediment. (b) Seaweeds growing over a coral substrate trap mud trapped by seaweeds. This mud is harmful to corals because it ultimately falls on the coral. Photos courtesy of L. McCook.

(a)

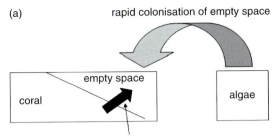

rapid colonisation of empty space

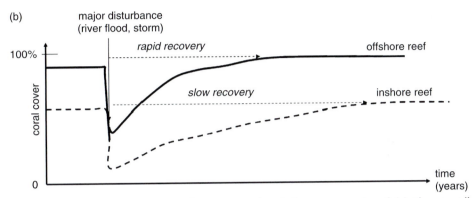

(b)

FIGURE 5.11. Sketch of the space war between coral and algae on a coral reef. (a) Algae rapidly colonise empty space made vacant by the death of corals following a major disturbance such as a river flood or a storm. Coral slowly recolonises that lost space by displacing the algae. (b) Coral cover decreases following such a major disturbance and the recovery period is longer in inshore than offshore reefs, because the quality of the water and the substrate is lower inshore than offshore.

on the water and substrate quality. This is because (1) suspended sediments cloud the water column and coats the substrate, thereby reducing photosynthesis, and (2) coral larvae recruit by attaching themselves to the reef and growing into adult colonies, and they are unable to do so in areas of high sedimentation or sediment buildup. Therefore, inshore reefs usually have higher coral cover than offshore reefs, located further away from the mouth of estuaries (Birkeland, 1997). The recovery period is shorter offshore than inshore due to higher water clarity (Fig. 5.11b).

The impact of the river inflow of freshwater and mud on coastal coral reefs is long lasting because of the long residence time of mud (Fig. 5.12). During the river flood, a river plume is formed that can impact directly on the reef. The suspended mud settles out, a fraction falls directly on the coral and smothers it. The remaining mud settles on the bottom around the reef to form a settled layer that can be either compacter or un-compacted (a nepheloid layer). This mud is resuspended during storms and decreases photosynthesis by reducing visibility. Some of that resuspended mud settles on the coral and harms it, and this occurs even without river runoff. As a result of this frequent sedimentation, fleshy and

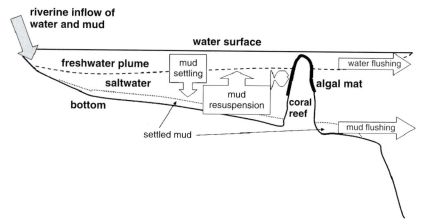

FIGURE 5.12. Sketch of the processes controlling the impact on coastal coral reefs from river inflow of freshwater and mud.

filamentous algae overgrow the coral and prevent coral recruitment (Fig. 5.11a). The impact disappears only after the mud is flushed away. The impact is long-lived because the residence time of mud is much larger than the residence time of water (Fig. 3.17b; Wolanski et al., 2003 a and b, 2004b and 2005; Golbuu et al., 2003; Victor et al., 2006). This can lead to an ecological phase shift, whereby the corals die and the substrate is entirely covered by algae, a process that is facilitated if nutrients are abundant (Done, 1992b).

CHAPTER 6

Ecohydrology models

For practical applications scientists and engineers are often called to quantify human impacts on estuarine ecosystems. This is usually carried out using numerical models. Because the physics often drive the biology, these models commonly start from models of the dynamics of water and sediment. The models of the physics usually perform satisfactorily because the equations of motion are known. However, a number of coefficients still need to be derived empirically from field observations and the open boundary conditions (e.g. what is happening in the adjoining ocean) are often poorly known (see chapters 2 and 3). Such models can be one-dimensional (usually along the estuary), two-dimensional (usually either a vertical slice along the estuary or a horizontal slice along the estuary), and three-dimensional. Model realism increases with an increasing number of dimensions at the cost, however, of increasing complexity and increasing skills needed to use the model. The model output is usually displayed, and the models are interrogated, along horizontal slices (e.g. Fig. 2.4) or vertical slices (e.g. Fig. 6.1). Thus models predict the physics of the water and the sediment throughout a tidal cycle for any forcing from the river (e.g. a flood or a drought), from the sea (e.g. tides and storm surge) and the wind.

To these models of the physics are attached sub-models describing the chemistry and biology of the estuarine ecosystem. Because the biology is extremely complex (see chapter 5), the combined physics-chemistry-biology model to reproduce in detail the ecosystem can become overwhelmingly complex and impractical. Thus it becomes necessary to simplify the estuarine ecosystem model by simplifying each of its components. Estuarine ecosystem modeling becomes a compromise between the need to increase complexity required for thoroughness and the need for practicality to simplify the sub-models to the most important components and processes.

Through the last 50 years engineers and scientists have diverged on the ways to find a compromise between practicality and realism. By and large this divergence still exists. Engineers are still focusing on refining sub-models of practical applications such as quantifying the fate of pollutants carried by water currents that are studied in great detail while the biology may be oversimplified (e.g. Price and Reed, 2006). Biologists are still focusing on food webs and ecosystem functioning while the physics may be oversimplified (e.g. Jorgensen et al., 1986;

may settle and rot in situ, thus locally depress the DO without being carried away by the currents. This model really assumes that the water is so grossly polluted that only bacteria thrive. Other aquatic biology processes occur that also assimilate the BOD. In practice in heavily eutrophicated water bodies with severe DO deficits there may be large day/night fluctuations in DO whereby algae and phytoplankton aerate the water column in daytime by photosynthesis and depress the DO at night by respiration (e.g. Mnaya et al., 2006). Fish may also be important in improving the DO by feeding on the solid waste. Finally

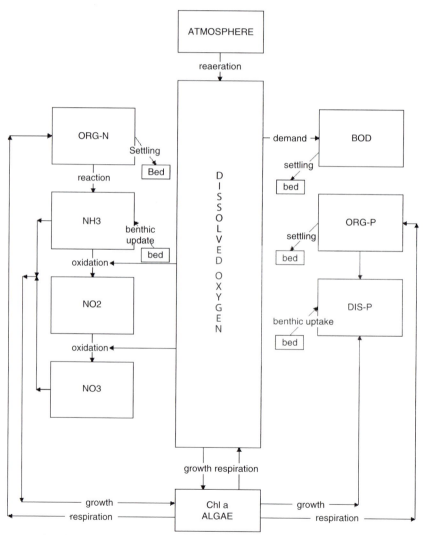

FIGURE 6.4. Sketch of the biology sub-model in the RMA-11 model. ORG = organic; DIS = dissolved. Chl a = chlorophyll a. Courtesy of I. King.

there may be interaction with the benthos whereby some of the waste may be removed by filter feeders such as bivalves.

To introduce more biology in models of wastewater impact on an estuary, engineers developed a suite of models to predict DO deficits. Probably one of the most advanced ones is the RMA-11 model. This model is used widely in pollution forecasting. It is used, for instance, to study the water quality of the whole of San Francisco Bay, U.S.A., in the presence of numerous sources of wastes and engineering works, to aid in planning remediation measures. The ecological sub-model of the RMA-11 model (Fig. 6.4) still assumes that the system is grossly polluted. It calculates the DO deficit. DO is influenced by aeration by the atmosphere, oxygen depletion by BOD, oxygen input by algal photosynthesis in daytime and oxygen removal by algal respiration in daytime. The growth of algae as a result of the waste discharge is modelled by considering the two nutrients nitrogen and phosphorus. Some of the organic phosphorus from the waste and excretion of the algae settles on the bed; the remaining fraction is converted by bacteria to dissolved phosphorus, which in turn is available for algae growth (see chapter 5). The organic nitrogen from the waste and algae excretion is converted to NH_3, which oxidizes to NO_2 and NO_3 in the presence of dissolved oxygen. This oxidization extracts more DO from the water column, and further depresses the DO. The model uses first order kinetics for oxidation.

There are several others similar models by large engineering consulting groups, such those of the DHI model in Denmark, the Wallingford model in the U.K., Delft3D in the Netherlands, and the MOHID model in Portugal. All these models essentially perform similar calculations with similar assumptions. Engineers commonly use their models to study one development scheme (e.g. a waste outfall or a construction that modifies the physics of an estuary) and they usually neglect all the other processes and developments impacting an estuary. The physics of the water is modelled in great detail; the dynamics of the fine sediment commonly neglect the biology ignoring a vast body of knowledge; the food web is usually greatly oversimplified. These engineering models still are targeted at highly polluted water bodies. They are not ecosystem models because they neglect the functioning of the ambient ecosystem. Specifically, they neglect wetlands and the natural populations of bacteria, plankton, fish, bivalves, shrimps, birds, and crabs. Thus they do not address the issue of the 'quality of life' that a wealthy population living on the shores of the estuary demands.

6.2. ECOSYSTEM MODELS

6.2.1. Predator-prey relationship

Scientific data are available on the functioning of an estuary as an ecosystem. This knowledge is still not integrated in working engineering models. Thus engineers and water resources managers lag the science. However science-based models of estuarine ecosystems are often impractical, in the sense that the number of parameters needed to run the model may be excessive and many of them may be unavailable. Many of these parameters necessitate further research.

Estuarine and coastal waters eutrophication models are many. The simplest ones describe spatially and temporally averaged ecosystems based on a few variables, while the most complex ones describe the spatial and temporal distribution of a large number of variables, including a limiting nutrient (Engqvist, 1996; Jorgensen et al., 1986). The most complex models are impractical to use because they have several hundreds of parameters (e.g. Fulton et al., 2004). All models require simplification to be practical. For instance, while there are several forms of phytoplankton and zooplankton (Fig. 5.2), an ecosystem model may put together in one 'phytoplankton' category all the size ranges of phytoplankton, and all the size ranges of zooplankton into one 'zooplankton' category.

There remains the difficult question of how to represent mathematically the prey-predator relationship between phytoplankton and zooplankton. The equation to use expresses the fact that mass is conserved. It is usually assumed to be of the Lotka-Volterra type although this form of equation itself is semi-empirical (May, 1974; Flint and Kamp-Nielsen, 1997; Hilborn and Mangel, 1997; Kot, 2001; Jorgensen and Bendoricchio, 2001). In its simplest form this equation takes the form,

$$d\,Phyto/dt = k_{phyto}\,Phyto \qquad (6.6)$$

where Phyto is the phytoplankton biomass and k_{phyto} is the phytoplankton growth rate. This leads to an exponential growth rate for phytoplankton if k_{phyto} is a constant (Fig. 6.5). However k_{phyto} is generally not a constant; it depends on the availability of the limiting nutrient and light. The limiting nutrient can be nitrogen, phosphorus or silicon (Fig. 5.8) and its depletion rate $k_{nutrient}$ is often calculated assuming that Michaelis-Menten kinetics prevail,

$$k_{phyto} = k_1\,N_{utrient}/(K_1 + N_{utrient}) \qquad (6.7)$$

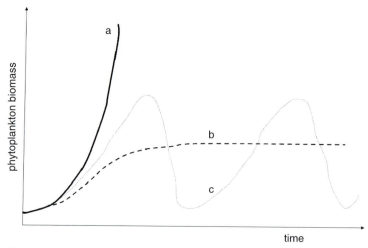

FIGURE 6.5. Time series of trajectories of the phytoplankton biomass for three mathematical scenarios resolved by the model: a: exponential growth; b: S-growth growth curve leading to a steady-state; c: instabilities develop as a series of population explosion and population crash.

where k_1 and K_1 are constants and $N_{utrient}$ is the concentration of the limiting nutrient. If the dissolved nutrient is limited, then its biomass is calculated by assuming that it is depleted at the same rate than it is used by phytoplankton while it recovers some nutrient from remineralisation of detritus R and from external input I,

$$d \, Nutrient/dt = -k_{phyto} \, Phyto + R + I \qquad (6.8)$$

If the phytoplankton is grazed by zooplankton, Eq. 6.6 is modified,

$$d \, Phyto/dt = k_{phyto} \, Phyto - G \qquad (6.9)$$

where G is the phytoplankton loss rate due to grazing by zooplankton. G is not a constant because it depends on both the phytoplankton biomass and the zooplankton biomass, both of which vary in time. A steady solution can be found whereby the phytoplankton biomass reaches a steady state after an S-growth curve (Fig. 6.5). Alternatively the system can be unstable with a succession of events of population explosion and population crash for both the prey and the predator (Fig. 6.5; May, 1974; Jorgensen and Bendoricchio, 2001). Thus, dependent on details of the mathematical formulation, there are at least three trajectories that the phytoplankton population can follow. The robustness of the model depends then on details of the equations and on the availability of a sound data set. The robustness of the model, or its failure, is thus in the details of the mathematical formulation and the availability, or otherwise, of a sound data set to verify the model performance for a wide range of situations including some that the model may not have been designed for (Hilborn and Mangel, 1997).

6.2.2. Estuarine ecosystem models

There are a large number of parameters required to formulate a simple food web. This makes models cumbersome to use because the data are often unavailable to accurately estimate these parameters. This is demonstrated below for the simple food web shown in Fig. 5.1 that assumes clear waters and no microbial loop.

Eq. 6.6 is used in which (Naithani et al., 2007),

$$k_{phyto} = \mu - r_p - m_p - w_s \qquad (6.10)$$

where μ is the daily phytoplankton biomass growth rate, r_p is the respiration, m_p is the natural phytoplankton mortality, and w_s is the settling rate. This equation is simple. It is in expressing the terms on the right hand side of Eq. (6.10) that the complexity arises.

The equation for μ is specific to a dominant phytoplankton species, and, if phosphorus is the limiting nutrient,

$$\mu = 2 P_{max} \, f(P) \, f(I) \qquad (6.11)$$

where P_{max} is the maximum photosynthetic rate, $f(I)$ is the light limitation factor and $f(P)$ is the phosphorus limitation factor. The equations for the two limitation factors are,

$$f(P) = Phos/(Phos + k_{phos}) \qquad (6.12)$$

$$f(I) = (1/k_e H)\{arctan[I_o/2 I_k] - arctan[I_o exp(-k_e H/2 I_k)]\} \qquad (6.13)$$

where k_{phos} is half saturation constant for phytoplankton, $k_e =$ the extinction coefficient, I_k is the light saturation constant, I_0 is the solar insolation reaching the surface of the water, and H is the water depth where phytoplankton occurs.

The loss rate is,

$$G = g\,Zoo \qquad (6.14)$$

where Zoo is the zooplankton biomass and g is the grazing rate,

- $g = g_{max}$, if Phyto $\rightarrow$ ILL $\qquad (6.15a)$
- $g = g_{max}\,(Phyto - Phyto_0)/(Phyto - Phyto_0 + \beta_{phyto})$, if Phyto $>$ Phyto$_0$ $\quad (6.15b)$
- $g = 0$ if Phyto $\leq$ Phyto$_0$ $\qquad (6.15c)$

where g_{max} is the maximum ingestion rate per day, ILL is the phytoplankton maximum value that is usually determined by light limitation, Phyto$_0$ is the phytoplankton minimum value below which zooplankton does not graze, and β_{phyto} is the phytoplankton half-saturation constant.

The value of r_p is determined empirically from measurements of the net production NP,

$$NP = 2\,P_{max}\,f(I)\,z - r_p\,Phyto\,H \qquad (6.16)$$

where z ($= 4.6/k_e$) is the euphotic depth.

There are thus 12 parameters to use in the phytoplankton model. Further parameters are also needed for the zooplankton model and for the fish model. For the zooplankton biomass, the continuity equation (6.7) becomes,

$$d\,Zoo/dt = k_{zoo}\,Zoo - p - m_z \qquad (6.17)$$

where k_{zoo} is the zooplankton gross growth rate, p is the predation loss, and m_z is the mortality rate of zooplankton. Settling is not included in the model, contrary to the case for phytoplankton (Eq. (6.9)) because zooplankton is mobile. A further 10 parameters are needed for the zooplankton model. A similar number of parameters are needed for the fish model. Therefore there is a need to determine about 30 parameters for the simple food web model shown in Fig. 5.1.

Some of these many parameters can be determined from laboratory or microcosm experiments. Other parameters can be deduced from fitting the model to observations (this is known as model calibration), and then running the model against another data set to validate the model. In most aquatic and marine waters, the data are insufficient to do all that. Hence to a large degree marine ecosystem models remain qualitative more than quantitative. Thus estuarine and coastal sea ecosystem modeling remains an art more than a science (e.g. Radach and Moll, 2006).

6.2.3. An estuarine ecohydrology model

In Eqs. (6.9)–(6.16), it is necessary to specify process-based parameters that essentially set the lower and upper limits of the population. A simpler formulation

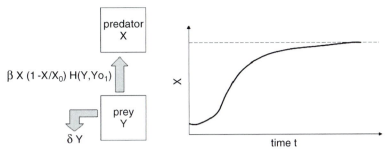

FIGURE 6.6. The simple prey-predator model described by Eq. (6.14) is based on starvation and saturation limits. It also readily accommodates the death δY from exceptional events such as when freshwater plankton dies and becomes detritus on being advected into the saline water region of the estuary. With unlimited prey, the predator biomass X reaches a steady state after an S-shape growth curve.

of the equations is to recognise a priori that such limits exist and are available from long-term, monitoring-type, field studies. It is then possible to write the prey-predator relationship to be governed by these limits,

$$dX/dt = \beta X(1 - X/X_o)\, H(Y, Y_{o1}) - \delta Y \qquad (6.18)$$

where (Fig. 6.6) X is biomass of the predator, Y is the biomass of the prey, β is the prey growth rate, X_o is the saturation biomass, Y_{o1} is the starvation biomass of the prey Y, δ is the death rate of the prey not due to natural birth/death processes or prey-predator relationships (e.g. the death of freshwater plankton when the currents bring them into saline water). The disadvantage of that method is that biological processes may be over-simplified. The advantage is that the number of parameters in the prey-predator relationship is decreased from 12 to 3. This greatly simplifies the model and allows the development of more complex, practical, food web models. For instance the simple food web shown in Fig. 5.1 using the process-based Eqs. (6.6–6.17) uses more parameters than the much more complex food web shown in Fig. 6.8 described using Eq. (6.18).

   This simpler ecosystem model is best suited to vertically well-mixed estuaries. It can readily be linked to a physical sub-model (Fig. 6.7) that views the estuary as a series of connecting cells. These cells exchange water by diffusion (tidal mixing) and by advection (the currents driven by the river runoff, the wind, and the oceanic inflow). The upstream cell receives freshwater, fine sediment with its particulate nutrients, dissolved nutrients, detritus, and freshwater plankton. The downstream cell receives seawater, clean marine sediment and detritus, dissolved nutrients and plankton. The physical processes and the open boundary conditions control the salinity distribution and the suspended solid concentration (SSC) in the estuary. An estuarine turbidity maximum zone (ETM) forms.

   This physical sub-model model is linked to an ecological sub-model. Within each cell of the model (Fig. 6.7), a food web exists that follows Fig. 5.5. This is further simplified to the food web model shown in Fig. 6.8 if the exchange

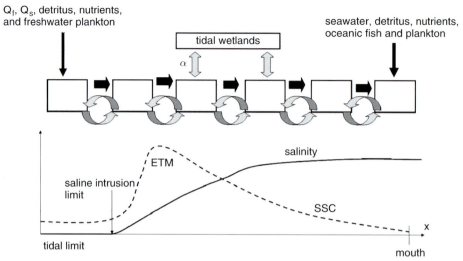

Q$_f$, Q$_s$, detritus, nutrients,
and freshwater plankton

seawater, detritus, nutrients,
oceanic fish and plankton

tidal wetlands

salinity

ETM

saline intrusion
limit

SSC

x

tidal limit

mouth

FIGURE 6.7. Sketch of the estuarine ecohydrology model. The physical sub-model divides the estuary into cells spread along the channel, with distance x, starting at the tidal limit and ending at the mouth. The upstream cell receives the riverine discharge of water (Q$_f$), sediment (Q$_s$), detritus, freshwater plankton and nutrients. The downstream cell receives oceanic water, detritus, nutrients, plankton and fish. There is a flux from cell to cell, unidirectional (black arrows) as a result of the freshwater discharge, and bidirectional (curved arrows) as a result of tidal mixing. An estuarine turbidity maximum (ETM) results, as well as a salinity intrusion limit. In every cell, a food web exists, such as that shown in Fig. 6.8 described later. There may also be a lateral import, or an export, in every cell at a rate α of material from/to tidal wetlands (e.g. a loss of fine sediment and a gain of fish and detritivores that took refuge as juvenile in tidal wetlands).

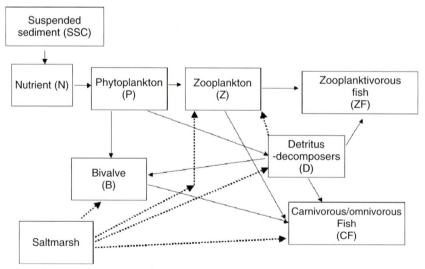

FIGURE 6.8. The food web in the muddy Guadiana Estuary, Portugal. Adapted from Wolanski et al. (2006b).

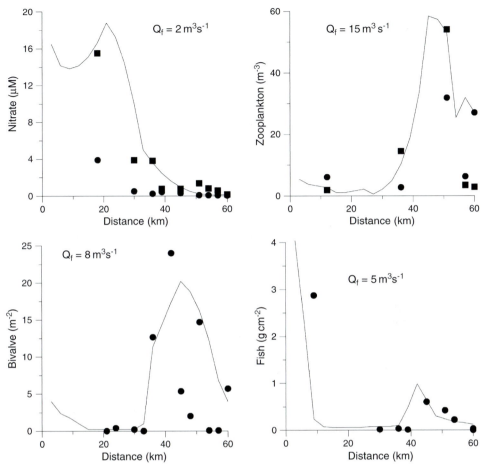

FIGURE 6.9. Observed (• and ■ for different years) and predicted (line) along-channel distribution in the Guadiana Estuary, Portugal, of nitrate, zooplankton, bivalve and fish during low flow conditions. Redrawn from Wolanski et al. (2006b).

of nutrients between the water and the substrate is negligible compared to that between the water and the wetlands, such as in the Guadiana Estuary, Portugal (Wolanski et al., 2006b).

Even though the model is simple, it is able to reproduce faithfully the observations on the distribution of nitrate concentration and biomass of fish, bivalve, nitrate and zooplankton during low flow periods (Fig. 6.9).

Important ecological processes that the model reproduces are (1) the long residence time of turbid waters enables the suspended solids to release particulate nutrients into the dissolved form and thus sustain the food web, (2) the role of detritus in supporting the food web, (3) the role of bivalves in decreasing phytoplankton; (4) the role of wetlands as a source of detritus as well as a nursery ground, and (5) river floods attract coastal fish to migrate up-estuary by kinesis or taxis by swimming following environmental clues, primarily salinity gradients. The model has been successfully verified against field data for the saltmarsh-fringed Guadiana Estuary, Portugal, and the mangrove-fringed Darwin Harbour estuary, Australia (Wolanski et al., 2006 b and c). For tropical Darwin Harbour estuary, the system has twice the level of complexity for phytoplankton and zooplankton than the temperate Guadiana Estuary.

It can also be used to model the onset of toxic algae blooms in an estuary following the biology sub-model shown in Fig. 5.8 (Chicharo et al., 2006).

Thus the ecohydrology model has the drawback to simplify, and possibly to over-simplify, prey-predator process. However, it has several advantages. Firstly it is mathematically stable because it incorporates an equal complexity at the top and bottom of the food web (Brauer and Castillo-Chavez, 2001). Secondly, it is simple enough that it can be used for practical applications, while other more complex estuarine models may be unwieldy as they comprise up to 12 state variables and over 50 parameters for which data are usually unavailable or insufficient (e.g. Flint and Kamp-Nielsen, 1997). Thirdly, as new data become available, the model can be made more complex, for instance by subdividing compartments (e.g. phytoplankton) into several classes. Finally the model takes a holistic viewpoint that considers the whole estuarine ecosystem including the riverine and oceanic influences (Fig. 1.9). It thus enables to link human land-use and water-use in the river catchment to water quality and ecosystem health in the estuary. The model naturally incorporates both bottom-up and top-down ecological controls.

Thus the model can be used for scenario testing. For instance it can be used to predict the impact on the estuary of irrigation farming from the Alqueva dam on the Guadiana River and of infilling the saltmarshes along the Guadiana Estuary for housing and golf courses. The model predicts that, as a result of irrigation farming, the phytoplankton concentration will double near the salinity intrusion limit as a result of top-down ecological control following the increase in the riverine nutrient discharge (Fig. 6.10). If the saltmarshes were destroyed, the phytpoplankton is also predicted to double in the estuary, this time near the upper limit of the fringing saltmarshes as a result of the decrease in bivalve biomass, showing the importance of bottom-down ecological control.

The model can also be used to assess the importance of short river floods (freshets) in maintaining biodiversity in the estuary. It suggests that in the absence of freshets the carnivorous fish population is minimal for 6 months a year (Fig. 6.11).

This finding demonstrates that the outflow from dams must be manipulated to generate freshests in dammed rivers in order to maintain the ecological health of estuaries.

The Guadiana Estuary sub-model described above is derived from the pelagic food web shown in Fig. 5.5. by neglecting the benthic food web. This model is further simplified for shallow, eutrophicated estuaries with extensive intertidal mudflats, such as the saltmarsh-fringed Ria Formosa, Portugal (Nobre et al.,

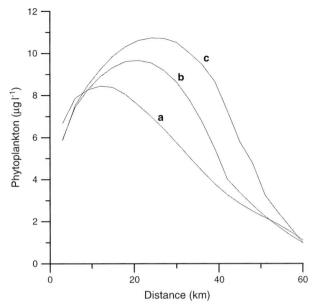

FIGURE 6.10. Along-channel distribution of predicted phytoplankton concentration in the Guadiana Estuary, Portugal, for the low-flow season (a) in the present conditions, (b) for a doubling of riverine nutrient concentration, and (c) for the removal of the saltmarshes. The distance starts at the tidal limit, km 60 = the river mouth. The increased phytoplankton concentration as a result of an increased riverine nutrient concentration is a result of bottom-up ecological control. The increased phytoplankton concentration as a result of the removal of the saltmarshes is a result of top-down ecological control because the removal of the wetlands would destroy the bivalves that prey on phytoplankton. Redrawn from Wolanski et al. (2006b).

2005). Due to eutrophication, the pelagic food web is much simplified (Fig. 6.12). The contribution from the benthos is dominant due to the presence of extensive macroalgae (seaweeds) covering the substrate, which causes a dissolved oxygen deficit at night from respiration. Saltmarshes, zooplankton and fish are neglected in the nutrient cycling. Similar equations as above are used to model such stressed systems.

Clearly thus, estuarine food webs vary from estuary to estuary. There is no unique model. However, once a food web model is developed for a particular estuary, an estuarine ecohydrology model can readily be developed using the simple prey-predator relationship of Eq. (6.18). The ecohydrology model incorporates the role of intertidal wetlands; these are saltmarshes in temperate estuaries and mangroves in tropical estuaries. Similar ecohydrology models can also be readily used to assess the impact of land-use activities on promoting toxic algae blooms; the same models can also be used to predict the usefulness of freshets from dams in reducing these blooms (Chicharo et al., 2006).

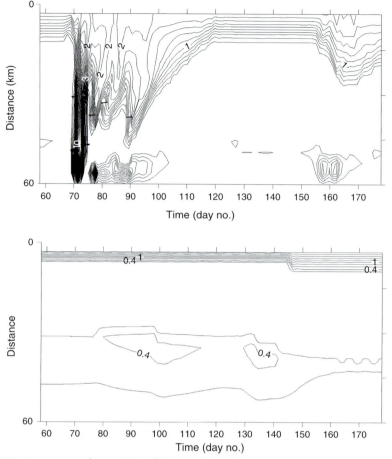

FIGURE 6.11. Time series plot in 2003 of the predicted distribution in the Guadiana Estuary, Portugal, of the carnivorous fish biomass (in g cm^{-2}) (top) without and (bottom) with the Alqueva dam that stopped any freshets. Redrawn from Wolanski et al. (2006b).

6.3. CORAL REEF ECOHYDROLOGY MODEL

Like the estuarine ecohydrology model described above, the coral reef ecohydrology model is also based on (1) linking physical and ecological sub-models, and (2) using Eq. (6.18) for prey-predator relationships (Wolanski et al., 2004b; Wolanski and De'ath, 2005).

The physical sub-model is sketched in Fig. 6.13a. It predicts (1) the intensity and duration of river plumes on coral reefs for individual years based on observed rainfall, (2) the yearly storm intensity for individual reefs, (3) the oceanographic connectivity between reefs, and (4) the turbidity during the coral recovery period. The sub-models (1) and (2) (i.e. river plumes and storms) yield yearly mortality of corals for individual reefs. The sub-model (3) (connectivity)

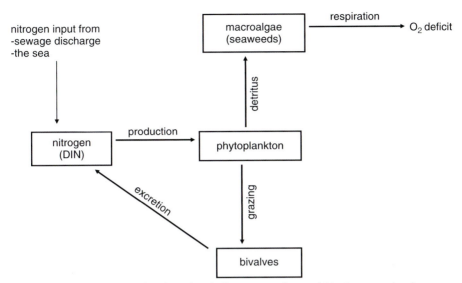

FIGURE 6.12. The pelagic food web in the shallow, eutrophicated Ria Formosa that has extensive intertidal mud banks covered with benthic macroalgae. Drawn from the equations of Nobre et al. (2005).

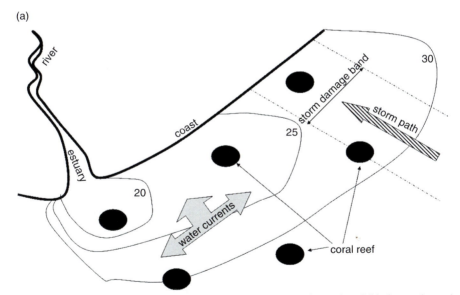

FIGURE 6.13. A sketch of the principal components in (a) the physical and (b) the ecological sub-models. SSC= suspended sediment concentration (turbidity). In (a) the contour lines are idealized surface salinity contours in parts per thousand. In (b) the thick lines show a transfer of biomass, the dotted lines show an ecological link (e.g. an increased SSC leads to higher turbidity, hence to lesser algal photosynthesis).

(b)

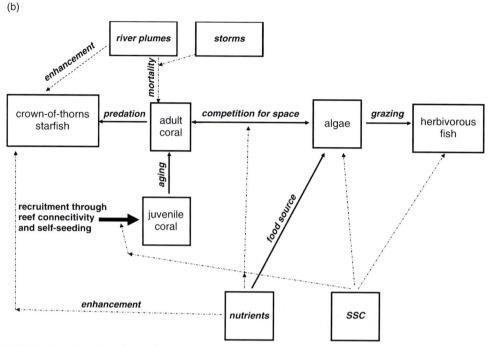

FIGURE 6.13. Continued

predicts the exchange of coral planulae between reefs following coral spawn-ing, and thus the yearly recruitment of coral juveniles that can re-populate an impacted reef.

The biological sub-model is sketched in Fig. 6.13b and based on Fig. 5.11a. It includes (1) the competition for space between the algae and the coral, (2) the prey-predator relationship between corals and crown-of-thorns starfish, and (3) the grazing of algae by herbivorous fish. The competition for space between corals and algae is also parameterized using Eq. (6.18). The suspended sediment concentration (i.e. the turbidity) and nutrients modulate all these processes.

The main human impacts are (1) the increased water turbidity as a result of land-use, (2) the degree of fishing, and (3) increased mortality from global warming.

This model was successfully verified for the 400 km-long central region of the Great Barrier Reef of Australia, which is the region most impacted by human activities on land (Fig. 6.14).

The model suggests that land-use has contributed to the degradation of the health of the reef (Table 6.1). This table also predicts that the health of the Great Barrier Reef will significantly worsen by the year 2050, and that the effect will be worsened by global warming because it results in an increased coral mortality. Table 6.1 also suggests that much-improved land-use practices will enable some regions of the Great Barrier Reef to recover even with global warming. Thus, an

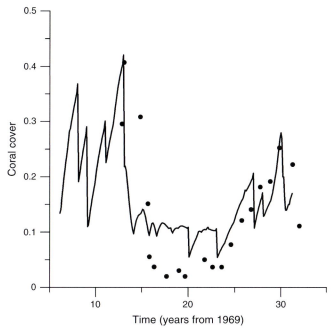

FIGURE 6.14. Time series plot of observed (•) and predicted (line) coral cover (0.5 = 50%) at John Brewer reef, a mid-shelf reef in the central region of the Great Barrier Reef, Australia. Modified from Wolanski and De'ath (2005).

TABLE 6.1. Ecohydrology model predictions of the average coral cover (0.5=50%) over 261 reefs in the 400 km-long central region of Australia's Great Barrier Reef for various scenarios.

Year	Average coral cover
Before European colonisation	0.67
1970	0.59
2005	0.49
2050: Scenario 1: without global warming and continuing present land-use practices	0.32
2050: Same as scenario 1, with 50% decrease in fine sediment and nutrient flux from land-use	0.52
2050: Scenario 2: same as scenario 1 with global warming (IPCC prediction A2).	0.28
2100: Scenario 2	0.1

ecohydrology solution is needed for the Great Barrier Reef in view of human activities in all its the river catchments, in the same manner that an ecohydrology solution is needed for estuaries world-wide in view of human activities in their individual river catchments.

time. On reaching adulthood these fish return to the river and the reservoir to help controlling HABs through top-down ecological control. As another example, plants and plant communities can be used to accelerate desirable hydrological and ecological processes, for instance in the use of riparian zones to trap nutrients and sediments. Riverine woodlands, most of which has been lost in lowland Europe and North America to intensive agriculture and urban development, can be used to regulate catchment-dependent fluxes (Peterken and Hughes, 1995). Floodplain woodlands can be created to reduce peak flood flow, maintain low flow, improve sediment and pollutant retention and nutrient sequestration, and increase habitat diversity and timber production (Kerr and Nisbet, 1996).

In freshwater, ecohydrology has been successfully applied to constructed wetlands, rivers and floodplains, for issues concerning aquaculture, managing shorelines and river beds to maximize fish yields, and improving water quality in reservoirs (UNESCO, 2006).

7.2. ESTUARIES

In the estuary it is generally not possible to use purely engineering solutions to solve environmental degradation problems. Generally estuarine environmental restoration can only be carried out in robust estuaries by restoring some of the ecological processes of the estuary (Fig. 1.9). This includes restoring or creating intertidal wetlands and combining this with a mix of engineering interventions. Engineering interventions include not only maintaining minimum environmental flows in dammed rivers but also generating freshets from dams, and diminishing the residence time of stressed estuaries by dredging to deepen channels to facilitate water flows, and by opening new river mouths.

For non-robust estuaries, even all these interventions may not work. Such is, for instance, the case of the eutrophicated Peel-Harvey Estuary, Australia, where a new channel was constructed to decrease the estuary residence time. This effort was not accompanied by land use remediation measures. As a result, many symptoms of eutrophication, such as algal blooms and large-scale fish kills, are still evident in the lower reaches of the river systems. In such cases of non-robust estuaries, a basin-wide ecohydrology solution is required that involves (1) manipulating the river ecosystem to decrease its impact on the estuary, and (2) regulating basin-wide human activities that impact on the river.

Wetlands, including mudflats, mangroves and saltmarshes, need to be maintained, restored or created, because of their ability to trap sediment and pollutants, to convert excess water-born nutrients into plant biomass, to provide habitats for demersal and pelagic species, and to help protect the coast from increased erosion following sea-level rise and sediment starvation from damming. In temperate climates when the above-ground plant biomass dies in saltmarshes in winter, this plant biomass may need to be harvested annually to remove nutrients.

A powerful ecohydrology tool is to generate freshets from dams to maintain the biodiversity of the estuary, reduce the occurrence of toxic algae blooms, and maintain species size and thus prey-predator relationships (chapter 6; Rey et al., 1991; Gillanders and Kingsford, 2002).

Ecohydrology offers also another solution against toxic algae, namely, using bivalves to filter and pelletize excess nutrients and plankton. Such is the case of urbanised San Francisco Bay, U.S.A. where eutrophication is inhibited because half of the water is filtered daily by Asiatic clam *Potamocorbula that* is an invasive species (Nichols et al., 1986; Cloern, 1996).

Ecohydrology also offers the solution to use macrophytes to avoid toxic algal blooms in rivers and lakes. Toxic algal blooms do not appear for P-PO4 $< 30\,\mu g\,l^{-1}$ (Dunne and Leopold, 1978). Reaching such a small concentration may be in practice unrealistic for many systems because it would involve changing land use and farming practices. However, it is possible to avoid toxic algal blooms by reducing P-PO4 concentrations to only $120\,\mu g\,l^{-1}$, provided macrophytes are present; thus the required changes in land use and farming practices are much smaller.

Shell beds, e.g. oyster shell beds in Delaware Bay (Powell et al., 2006), must be managed not just for the economy but also for the environmental services that they provide. Shell beds are an essential fish habitat of many estuaries. They disappear when the shells are over-harvested and the beds are damaged by dredging. Management of estuarine fisheries must include management of the shell beds to ensure habitat maintenance.

The choice of measures varies from estuary to estuary. It is facilitated by using ecohydrology models for scenario testing.

7.3. COASTAL WATERS

Once eutrophication occurs in coastal waters, the situation cannot be redressed by local engineering solutions. The scale of coastal waters and their open boundaries makes a local management approach impossible. The problem is immense. Examples of this are the dead zones of hypoxic and anoxic waters in the Gulf of Mexico, the Baltic and Black seas. The reason is excess fertilizers discharged by rivers in coastal waters and originating from land-use activities.

For eutrophicated coastal waters, the only option available to restore environmental health is to adopt a basin-wide ecohydrology solution by managing human activities throughout the river catchment. One such management tool that should be required in the future for dammed rivers is the generation of occasional, man-made, river floods from dams. These floods are ecologically important because they "feed" the coastal waters with sediments and nutrients and they provide the adequate nutrients Si:N:P ratios to promote productivity. There is increasing evidence that coastal fisheries landings are related to the high river flows events and not climate factors, as has been demonstrated for South Portugal, the East Mediterranean coast, the Black Sea, and the Gulf of Carpentaria. For instance river floods in the Guadiana Estuary, Portugal, promote fish catches in coastal waters of anchovy (*Engraulis encrasicolus*) and sardine (*Sardina pilchardus*). This is because river plumes, through their high nutrient and organic matter loads, promote primary and bacterial production, increase the survival of larval and juvenile fish by increasing turbidity and reducing predation, and provide an environmental clue (salinity gradients) for shrimps and fish larvae to migrate towards the estuary where they will use the tidal wetlands as a nursery.

For coral reefs, science offers to economists and politicians the knowledge necessary to develop management policies that integrate socio-economics, land-use activities, and coral reef ecosystem health. In practice, however, for the Great Barrier Reef and most corals reefs worldwide outside of a few islands in Micronesia, science has little impact because of two phenomena. Firstly there is the "tragedy of the commons" where few take any responsibility but everyone has ownership – this is the same problem which has resulted in the collapse of fisheries worldwide. Secondly there is uncertainty in the science of cause and effects of reef degradation – some of that uncertainty is inherent to ecosystem science (see chapters 5 and 6), much is purposely manufactured in non-peer reviewed publications. This uncertainty enables politicians and other decision makers to ignore the problem and to implement no remedial measures on land use.

7.4. MANAGING HUMAN HEALTH THREATS

Wetlands must be managed to reduce human health problems. The issue of human health is critical when restoring, creating and managing wetlands, to minimize diseases (e.g. bilharzia and malaria) and vector breeding grounds.

When managing wetlands, it is important to recognize that malaria is a significant threat. Malaria affects 300 million people per year, and kills about 1 million per year. It is often erroneously thought that larvae of malaria-transmitting mosquitoes can only develop in freshwater. Malaria is not restricted to the tropics. About 200 years ago malaria was the leading cause of death in people living near saltmarshes in SE England in the Netherlands (Reiter, 2000). Only in 1975 did the World Health Organization declare Europe to be malaria free. Historically, malaria declined in North Europe due to drainage and land claim of saltmarshes. It was finally eradicated through the use of DDT (Reiter 2000). DDT was banned later on as it is persistent in the ecosystem and it was even found in women's breast milk. Non-persistent DDT alternatives were developed and used. Malaria is now resurging due to urbanisation creating stagnant water and to more mosquito resistance to DDT alternatives and to medication.

There is thus a possible conflict in third word countries between combating malaria (human health issue) and creating wetlands. However ill-planned urbanisation and non-planned urbanization (i.e. slums) have stagnant water holes and are a much more serious malaria health risk.

With a few exceptions described below, a rule of thumb in creating or managing wetlands is avoiding slow-flowing or stagnant waters, particularly under a scenario of increasing temperature which allows disease-causing organisms to extend their range to colder climates (WHO, 2000). For tidal wetlands, it is possible to do that by designing the tidal drainage pattern to avoid a long residence time; this must be less than 5 days for some malaria or Ross River vector-carrying mosquito larvae (Dale, 1993). The tidal drainage channel must be designed carefully to minimize disturbing acid sulfate soils because the environment consequences are severe and include fish kills (Fig. 1.5f; Soukup and Portnoy, 1986). The environmentally friendly yet effective technique in solving the mosquito problem is to drain the mosquito larval pools through a runnel ($< 0.3\,m$ deep, $0.9\,m$ wide spoon-shaped channel) to a tidal creek so as to allow tidal flushing and access to predators of the mosquito larvae (i.e. top-down ecological control; Dale and Knight, 2006).

The strategy is different for controlling the saltmarsh mosquitoes *Ochlerotatus taeniorrhynchus* and *Ochlerotatus sollicitans*. These will not lay their eggs upon standing water. To control them mosquito control impoundments are used (Clements and Rogers, 1964). The impoundment is a saltmarsh or mangrove forest with an earthen dike around the perimeter. To minimize damage to vegetation due to excessively long inundation, the impoundment is kept flooded only during the breeding season. This prevents oviposition. During the remaining part of the year, water levels inside the impoundment are allowed to fluctuate with the tides using culverts installed through the impoundment dikes. This eliminates mosquito production from the area without having to use pesticides. At the same time, the reconnection of the mosquito impoundment re-establishes the plankton communities and allows access to the marsh during most of the year to transient fish species (Rey et al., 1991; Brockmeyer et al. 1997).

7.5. HABITAT CREATION

Freshwater wetlands are used worldwide to improve the health of rivers and lakes by using existing wetlands and creating new wetlands to treat wastewater. There is now a large drive to do so also in saltwater wetlands.

7.5.1. Saltmarshes

7.5.1.1. U.S.A.

The U.S.A. has a national commitment to a goal of no overall net loss (NNL) of wetlands acres and functions, to be followed by a net gain, a goal that is incorporated in a free market economy using wetlands credit sales (Shabman and Scodari, 2004). The NNL goal is managed by Section 404 of the Clean Water Act (CWA) is administered by the U.S. Army Corps of Engineers (Corps) with oversight by the U.S. Environmental Protection Agency (EPA) (Strand, 1997; GAO, 2001). Anyone wishing to place fill material in a wetland that falls under the legal jurisdiction of Section 404 is required to secure a permit from the Corps. If a permit is issued, the developer has the responsibility to restore wetlands or create new wetlands to support the NNL goal.

To create saltmarshes, dredged sediment is pumped hydraulically and placed in shallow areas (Fig. 7.1; Streever, 2000). In the presence of waves, small breakwaters are built to protect against erosion and let the sediment dewater and

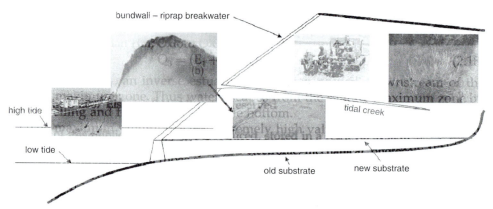

FIGURE 7.1. Creation of a saltmarsh using dredged sediment.

consolidate. Tidal creeks are dredged. *Spartina alternifora* grass is then planted mechanically, after being harvested from natural areas or grown in a nursery. Generally these created wetlands do not replicate the biological productivity of natural wetlands. However such created wetlands are still much preferable than sea disposal of the dredged mud and the pollution it creates.

To avoid the invasion by *Phragmites australis* into the upper reaches of salt-marshes, the traditional technique was repeated herbicide sprays followed by burning. A new technique involves blocking the invasion routes by planting desired plants selected from wild populations as well as tissue culture regenerants (Wang et al., 2006).

7.5.1.2. U.K.

Tidal wetlands are created by managed retreat (Fig. 7.2; Hazelden and Boorman, 2001; French, 2006). Old seawalls that were constructed centuries ago to con-vert saltmarshes into polders and farmland, are breached. Smaller seawalls are built further inland. A new saltmarsh is then created in front of them. The new saltmarsh dissipates wave energy and provides a protective buffer for the sea walls that may be built smaller. This created saltmarsh also pro-vides ecological services, such as organic matter and a refuge for fish and birds.

There are still difficulties in establishing the plant vegetation, a process that is facilitated by creating a suitable tidal drainage creek network before breaching the dykes.

7.5.1.3. Australia

Experience in restoring saltmarshes has shown that the process can be speeded up by transplantation saltmarsh plants from donor sites as well as using plants prop-agated in green houses (Laegdsgaard, 2006). Weeding and fencing a degraded saltmarsh also speed up the natural recovery process.

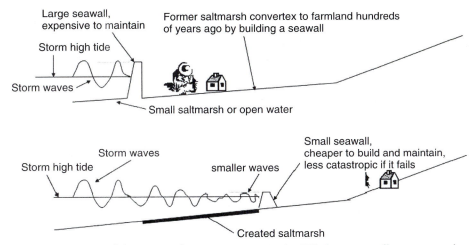

FIGURE 7.2. As part of the managed retreat strategy in the U.K., large seawalls constructed several centuries ago to reclaim saltmarshes for farmland are breached. Smaller seawalls are erected landward and are protected by a created saltmarsh.

7.5.2. Mangroves

The U.S. technique of creating saltmarshes using dredged mud is little used for mangroves. The only exception may be the recent project at Port Point Lisas, Trinidad, where in 2000, a 20 ha, shallow, mud bank was bunded, filled by dredge spoil, and allowed to dewater and consolidate. The bund wall was removed and the mud flat was naturally recolonised by mangrove seedlings from a nearby natural mangrove forest. There are, however, no published reports on the sustainability of this mangrove forest.

To protect the muddy coastline from coastal erosion caused by typhoons in Vietnam, there is a thriving, long-term and successful effort to plant mangrove trees in shallow, muddy areas along the coast (Fig. 4.6; Mazda et al., 1997b and 2006). To minimize uprooting of the mangrove seedlings by waves, the seedlings are grown in nursery and re-planted on site when the tree is at least 0.5–1 m tall. Using such mature seedlings grown in a nursery is more efficient than planting seeds or very young seedlings taken from the wild because crab predation on very young seedlings can reach 100% at some sites, though it is more commonly 70–80%.

Along canals and tidal creeks near urban areas, planted mangrove seedlings often fail to establish if they are pushed over by floating debris and boat-induced waves. A solution is used in Florida waterways. Each mangrove seedling is protected with a thin tube wall of polyvinyl chloride (Riley and Kent, 1999). The pipe is partially split with a thin blade in order to maintain sediment levels inside the tube while the growth of the tree is not restricted, and the tube is sufficiently rigid to be driven into and anchor itself in the soft bottom sediments.

Mangroves can be readily re-established in abandoned shrimp ponds where the substrate is well above mean sea level, simply by planting mangrove seedlings and breaching the levees to restore tidal flushing.

No reliable technique has yet been developed to economically and successfully replant mangroves in shrimp ponds excavated in mangrove soils. This is due to problems with acid soil leachate and contamination by viruses that usually result in a pond economic lifetime of only 5–8 years (Fig. 1.5d). Further, the pond bed level is usually near low tide level, which condemns the planted seedlings to drown. The soil needs to be entirely rebuilt to fill the hole and prevent acid leachate from exposed mangrove soils. Remediation is technologically feasible but economically too expensive for widespread use.

No technique has yet been developed and proven to successfully replant mangroves over oiled areas. This is because the oil has infiltrated the ground through crab holes where it can persist there for 10–20 years (Fig. 1.7 a and b).

7.5.3. Seagrass

The technology has been developed in Western Australia to plant seagrass in large quantities. A large underwater tractor is driven along the bottom over healthy seagrass (*Posidonia coriacea* and *Amphibolis griffithii*) beds (Paling et al., 2001). Large pods of undisturbed sediment with the seagrass are cut mechanically, placed in containers, and driven to the planting areas where they are planted with the original soil around them. The mean survival rate is 70% over 3 years in clear waters.

In Long Island Sound, U.S.A., eelgrass is being replanted manually by divers in a checkboard pattern consisting of squares 0.5 m by 0.5 m, at 50 shoots per square, alternating between unplanted and planted quadrats. Shoots were harvested from natural, healthy donor beds and tied to jute sheets stretched between PVC pipes. These beds are laid on the bottom.

Replanting seagrass in areas with occasional high wave energy has been a failure with zero survival after two years.

7.5.4. Coral reefs

Restoration efforts are largely restricted to the U.S.A. where there has been a great deal of efforts to restore coral reefs within and adjacent to estuaries, with limited positive results (Richmond, 2005). The transplantation of corals from a healthy reef to a damaged one always failed when the underlying causes of coral stress (i.e. poor land use) were not rectified. Failure also occurred when corals acclimated to one type of physical environment were transplanted to a different one. In a transplantation effort off Maui, Hawaii, 100% mortality of corals occurred over a 6-year period due to shifting sands and sedimentation.

Restoration activities that had a degree of success consisted of stabilizing damaged corals and rubble, "rebuilding" topographic relief by moving large coral heads and dislodged reef material into areas scraped clean by a ship, and transplanting both hard and soft corals to the site.

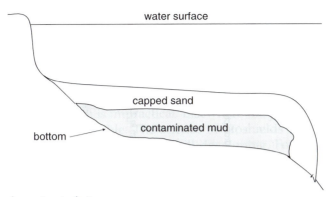

FIGURE 7.3. Sand capping technique.

The most effective efforts at coral reef restoration to date have been those that have focused on restoring those conditions of good land-use that allow natural recovery to occur.

7.5.5. Sediment capping

Estuarine sediment can remain polluted by heavy metals and synthetic chemicals for decades after the contaminant source has stopped. A remediation measure popular in Japan is to cover the contaminated material by a 0.5–1 m thick sand cap (Fig. 7.3; Furukawa and Okada, 2006). This decreases the re-entrainment of the pollutants in the water column and provides clean sediment as a habitat for organisms. This remediation measure lasts the longest in calm waters of semi-enclosed bays; it breaks down when the sediment cap is eroded or mixed with the contaminated material.

7.6. PROTECTION AGAINST NATURAL HAZARDS

Bioshields are areas of vegetation, including wetlands, which protect the coast, human lives and capital (Badola and Hussain, 2005; Prasetya, 2007; Preuss, 2007; Wolanski, 2007). The vegetation is usually planted or enhanced on hill slopes, river banks, along estuaries and coasts. The vegetation on steep hills protects against mud slides the human population living near the coast (Fig. 1.2d). Bioshields are also used to restore fisheries habitats in a degraded but sheltered estuary, protect an open coast against erosion, and protect the coast against a moderate typhoon and a moderate tsunami (Fig. 7.4). At the same time they provide ecological services that have socio-economic benefits to the human population, including timber and fisheries and lesser salt spray.

The protection that bioshields offer has limits in extreme natural hazards. It becomes a matter of living within accepted risks. The level of risk will vary from site to site with details of the bathymetry of the coastal waters and the topography of the coastal areas, the geology, the meteorology, and the oceanography.

dam operators. It is a far more stringent requirement than the existing environmental requirements worldwide, including the much lauded EU Water Framework Directive laws that set minimum environmental flows in dammed rivers. These laws are still ecologically flawed because there is no requirement to generate freshets.

Human history and ecohydrology science demonstrate that sustainable development requires changing our planning process to a watershed-wide, ecosystem level of managing the river catchment including the estuary and coastal waters. This requires to steer the political system to adopt ecohydrology to manage rivers, estuaries and coastal waters, instead of the present management principles based on regulating specific activities (e.g. farming, water resources, fisheries, urban developments), or based on political boundaries (cities, districts or counties) that are not those of the watershed ecosystem. Because these human practices have, throughout history, not recognized that estuaries and coastal waters are part of the river catchment, they have invariably failed the estuaries and coastal waters. The ecohydrology approach requires a high level of collaboration amongst stakeholders.

Worldwide the implementation of an ecohydrology-based strategy will keep on stalling until a political solution is found to regulate human activities on land. Indeed farmers, cattlemen, fishermen, and urban developers are often at odds with the imposition of ecohydrology-based rules for land-use, water resources management and fisheries, rules that they claim jeopardize their ability to earn a living. The implementation of the ecohydrology solution is however possible if it involves only one country dealing with its own land, freshwater, estuaries and coastal seas, because it becomes an issue of a national will and leadership. When that political will exists and is financially supported, human activities in the catchment may be regulated slowly, step-by-step, which is a complex socio-economic process, such as is attempted for Chesapeake Bay and for the Everglades, U.S.A. (U.S. EPA, 2001; U.S. EPA, 2002).

The partial restoration of some ecological functions in the Rhine and Thames estuaries is due to solving point discharges of pollution, without addressing land-use issues; this technological fix has not restored the full ecological health of the estuaries. In the Mersey River, U.K., as well as a number of examples in Europe, North America and Australia, a basin-wide approach is slowly taking shape in environmental management, but in all those cases the estuary and coastal waters are not considered explicitly in the objectives.

Legislative efforts in Europe are starting to recognise estuaries as being part of the river and river catchment ecosystem. Indeed, recent European Community directives address not only specific activities such as those related to the titanium dioxide industry, bathing waters' quality, urban waste water treatment, shellfish growing waters and shellfish hygiene, they also address habitats (*the Habitats Directive*), birds (the *Wild Birds Directive*), environmental impacts (the *Environmental Impact Assessment* and the *Strategic Environmental Assessment Directives*), pollution (the *Integrated Pollution Prevention and Control, the Environmental Liability, and the Nitrates Directives*), and ecological health (the *Water Framework Marine Strategy* and *the proposed Marine Strategy Framework directives*; Elliott et al.,

2006). This is a great step forward. It is still a long way from recovering the ecosystem health and biodiversity because there is no requirement to include estuaries in mitigating the environmental impact of dams, nor is there mention of the need to generate man-made freshets and even floods in dammed rivers.

Along the Asia Pacific coast including Australia and with the recent, partial exception of Japan – probably too late, however, to enable to recover estuarine ecosystems - the concept of ecohydrology is still an academic exercise that is not implemented in practice. In those countries, the degradation of estuaries and coastal waters is commonly accepted as a normal consequence of economic development (Wolanski, 2006a).

When the issue is international or global, the ecohydrology solution is much more difficult to implement because it requires a multinational commitment. This issue particularly affects coastal coral reefs for which the threat of global warming is most significant (Hughes et al., 2003; Hoegh-Guldberg, 2004). If global warming proceeds unchecked only biological adaptation could prevent a collapse of the Great Barrier Reef health by the year 2100 (Table 6.1). There are no data available on whether this hoped-for biological adaptation may or may not happen (Wolanski and De'ath, 2005). It may be that for coastal coral reefs worldwide there is no other politically acceptable solution at the moment than "do nothing about land-use and global warming and hope for the best".

Boorman, L.A., Hazelden, J.H., Loveland, P.J., Wells, J.G., 1994a. Comparative relationships between primary productivity and organic and nutrient fluxes in four salt marshes. pp. 181–189 in Mitsch, W.J. (ed.), Global Wetlands. Old World and New. Elsevier, Amsterdam.

Boorman, L.A., Hazelden, J., Andrews, R., Wells, J.G., 1994b. Organic and nutrient fluxes in four north-west European salt marshes. pp. 243–248 in Dyer, K.R., Orth, R.J. (eds), Changes in Fluxes in Estuaries: Implications from Science to Management. Olsen and Olsen, Fredensborg.

Boto, K.G., Wellington, J.T., 1983. Phosphorus and nitrogen nutritional status of a northern Australian mangrove forest. Marine Ecology Progress Series 11, 63–69.

Boto, K.G., Wellington, J.T., 1984. Soil characteristics and nutrient status in a northern Australian mangrove forest. Estuaries 7, 61–69.

Boto, K.G., Alongi, D.M.M., Nott, A.L. 1989. Dissolved organic carbon–bacteria interactions at sediment water interface in a tropical mangrove system. Marine Ecology Progress Series 51, 243–251.

Bourke, L., Selig, E., Spalding, M., 2002. Reefs at Risk in Southeast Asia. World Resources Institute, Cambridge, 72 pp.

Bowers, D.G., Binding, C.E., 2006. The optical properties of mineral suspended particles: a review and synthesis. Estuarine, Coastal and Shelf Science 67, 219–230.

Bowers, D.G., Harker, G.E.L, Smith, P.S.D., Tett, P., 2000. Optical properties of a region of freshwater influence (The Clyde Sea). Estuarine, Coastal and Shelf Science 50, 717–726.

Bowers, D.G., Evans, D., Thomas, D.N., Ellis, K., le B. Williams, P.J., 2004. Interpreting the colour of an estuary. Estuarine, Coastal and Shelf Science 50, 13–20.

Brauer, F., Castillo-Chavez, C., 2001. Mathematical Models in Population Biology and Epidemiology. Springer, Berlin, 416 pp.

Bridges, P.H., Leeder, M.R., 1977. Sedimentary model for intertidal mudflat channels with examples from the Solway Firth, Scotland. Sedimentology 23, 533–552.

Brinkman, R., Wolanski, E., Spagnol, S., 2004. Field and model studies of the nepheloid layer in coastal waters of the Great Barrier reef, Australia. pp. 225–229 in Jirka, G.H., Uijttewaal, W.S.J. (eds), Shallow Flows. Balkema Publishers, Leiden.

Brockmeyer, R.E., Rey, J.R., Virnstein, R.W., Gilmore, R.G., Earnest, L., 1997. Rehabilitation of impounded estuarine wetlands by hydrologic reconnection to the Indian River Lagoon, Florida. Wetlands Ecology and Management 4, 93–109.

Brush, G.S., Martin, E.A., DeFries, R.S., Rice, C.A., 1982. Comparisons of ^{210}Pb and pollen methods for determining rates of estuarine sediment accumulation. Quaternary Research 18, 196–217.

Burke, R.W., Stolzenbach, K.H., 1983. Free surface flows through salt marsh grass. Massachusetts Institute of Technology, Sea Grant College Program. Publication No. MITSG 83-16, Cambridge, Massachusetts, 252 pp.

Cahoon, D.R., Lynch, J.C., Powell, A.N., 1996. Marsh vertical accretion in a Southern California estuary. Estuarine, Coastal and Shelf Science 43, 19–32.

Carlson, P.R., Yarbro, L.A., Zimmermann, C.F., Montgomery, J.R., 1983. Pore water chemistry of an overwash mangrove island. pp. 239–250 in Taylor W.K., Whittier, H.O. (eds), Future of the Indian River System, First Symposium, Melbourne, Florida.

Cai, W.J., Wang, W.C., Krest, J., Moore, J.S., 2003. The geochemistry of dissolved inorganic carbon in a surficial groundwater aquifer in North Inlet, South Carolina, and the carbon fluxes to the coastal ocean. Geochimica et Cosmochemica Acta 67, 631–639.

Camenen, B., Larson, M., 2005. A general formula for non-cohesive bed load sediment transport. Estuarine, Coastal and Shelf Science 63, 249–260.

Chambers, R.M., Osgood, D.T., Kalapasev, N., 2003. Hydrological and chemical control of *Phragmites* growth in tidal marshes of SW Connecticut, USA. Marine Ecology Progress Series 239, 83–91.

Chanson, H., 1999. The Hydraulics of Open Channel Flows: An Introduction. Butterworth-Heinemann, Oxford, 512 pp.

Chappell, J. 1993. Contrasting Holocene sedimentary geologies of lower Daly River, northern Australia, and lower Sepik-Ramu, Paua New Guinea. Sedimentary Geology 83, 339–358.

Chen, S.-N., Sanford, L.P., Koch, E.W., Shi, F., North, E.W., 2006. A Nearshore Model to Investigate the Effects of Seagrass Bed Geometry on Wave Attenuation and Suspended Sediment Transport. Horn Point Laboratory, University of Maryland.

Chicharo, L., Chicharo, M.A., Ben-Hamadou, R., 2006. Use of a hydrotechnical infrastructure (Alqueva Dam) to regulate planktonic assemblages in the Guadiana estuary: basis for sustainable water and ecosystem services management. Estuarine, Coastal and Shelf Science 70, 3–18.

Chicharo, L., Chicharo, M.A., Esteves, E., Andrade, P., Morais, P., 2002. Effects of alterations in fresh water supply on the abundance and distribution of Engraulis encrasicolus in the Guadiana Estuary and adjacent coastal areas of south Portugal. Ecohydrology and Hydrobiology 1, 195–200.

Chong, V.C., Sasekumar, A., Wolanski, E., 1996. The role of mangroves in retaining penaeid prawns larvae in Klang Strait, Malaysia. Mangroves and Salt Marshes 1, 11–22.

Chong, V.C., King, B., Wolanski, E., 2005. Physical features and hydrography. pp. 1–16 in Sasekumar, A., Chong, V.C. (eds), Ecology of Klang Strait. Faculty of Science, University of Malaya, Kuala Lumpur, 269 pp.

Christie, M.C., Dyer, K.R., Blanchard, G., Cramp, A., Mitchener, H.J., Paterson, D.M., 2000. Temporal and spatial distributions of moisture and organic contents across a macro-tidal mudflat. Continental Shelf Research 20, 1219–1241.

Clements, B.W., Rogers, A.J., 1964. Studies of impounding for the control of salt marsh mosquitoes in Florida, 1958–1963. Mosquito News 24, 265–276.

Cloern, J.E., 1996. Phytoplankton bloom dynamics in coastal ecosystems – a review with some general lessons from sustained investigation of San Francisco Bay (California, USA). Review of Geophysics 43, 127–168.

Corbett, D.R., Dillon, K., Burnett, W., Schaefer, G., 2002. The spatial variability of nitrogen and phosphorous concentration in a sand aquifer influenced by onsite sewage treatment and disposal systems: a case study on St George Island, Florida. Environmental Pollution 117, 337–345.

Costa, M.J., Lopes, M.T., Domingos, I.M., de Almeida, P.R., Costa, J.L., 1996. Final report of the Portuguese Fauna Working Group – Tagus and Mira sites. pp. 95–174 in Lefeuvre, J.C. (ed.), Effects of Environmental Change on European Salt Marshes. Structure, Functioning and Exchange Potentialities with Marine Coastal Waters, Vol. 3. University of Rennes, France.

Dale, P.E.R., 1993. Australian wetlands and mosquito control – contain the pest and sustain the environment. Wetlands (Australia) 12, 281–307.

Dale, P.E.R., Knight, J.M., 2006. Managing salt marshes for mosquito control: impacts of runnelling, open marsh water management and grid ditching in sub-tropical Australia. Wetlands Ecology and Management 14, 211–220.

d'Alpaos, A., Lanzoni, S., Mudd, S.M., Fagherazzi, S., 2006. Modeling the influence of hydroperiod and vegetation on the cross-sectional formation of tidal channels. Estuarine, Coastal and Shelf Science 69, 311–324.

Dalrymple, R.W., Zaitlin, B.A., Boyd, R., 1992. Estuarine facies models; conceptual basis and stratigraphic implications. Journal of Sedimentary Petrology 62, 1130–1146.

Davies, A.J., Johnson, M.P., 2006. Coastline configuration disrupts the effects of large-scale climatic forcing, leading to divergent temporal trends in wave exposure. Estuarine Coastal and Shelf Science, 69, 643–648.

Day, J.W. Jr, Hall, C.A., Kemp, W.M., Yanez-Aranacibia, A., 1989. Estuarine Ecology. Wiley, New York, 558 pp.

de Castro, M., Gomez-Gesteira, M., Alvarez, I., Prego, R., 2004. Negative estuarine circulation in the Ria of Pontevedra (NW Spain). Estuarine, Coastal and Shelf Science 60, 301–312.

de Castro, M., Dale, A.W., Gomez-Gesteira, M., Prego,R., Alvarez, I., 2006. Hydrographic and atmospheric analysis of an autumnal upwelling event in the Ria of Vigo (NW Iberian Peninsula). Estuarine, Coastal and Shelf Science 68, 529–537.

de Graaf, G.J., Xuan, T.T., 1998. Extensive shrimp farming, mangrove clearance and marine fisheries in the southern provinces of Vietnam. Mangroves and Salt Marshes 2, 159–166.

Deleersnijder, E., Campin, J.-M., Delhez, E.J.M., 2001. The concept of age in marine modeling. I. Theory and preliminary model results. Journal of Marine Systems 28, 229–267.

Dermuren, A.V., Rodi, W., 1986. Circulation of flow and pollutant dispersion in meandering channels. Journal of Fluid Mechanics 172, 63–92.

De Villiers, M., 2000. Water: The Fate of our Most Precious Resource. Houghton Mifflin Company, Boston, MA, 352 pp.

Guan, W.B., Wolanski, E., Dong, L.X., 1998. Cohesive sediment transport in the Jiaojiang River estuary, China. Estuarine, Coastal and Shelf Science 46, 861–871.

Gueune, Y., Winett, G., 1994. The transport of the pesticide atrazine from the fresh water of the wetlands of Brittany to the salt water of the Bay of Mont Sainte-Michel (France). Journal of Environmental Science and Health 29, 753–768.

Hanley, R., 2006. Technical Guidelines for Reforestation of Mangroves and Coastal Forest in Aceh Province. Report to FAO, Rome, 55 pp.

Hannan, J.C., Williams, R.J., 1998. Recruitment of juvenile marine fishes to seagrass habitat in a temperate Australian estuary. Estuaries 21, 29–51.

Harrison, W.G., 1992. Regeneration of nutrients. pp. 385–406 in Falkowski, P.G., Woodhead, A.D. (eds), Primary Productivity and Biogeochemical Cycles in the Sea. Plenum Press, New York.

Hazelden, J., Boorman, L.A., 2001. Soils and managed retreat in South England. Soil Use and Management 17, 150–154.

Hazelden, J., Jarvis, M.G., 1979. Age and significance of alluvium in the Windrush valley, Oxfordshire. Nature 282, 291–292.

Heip, C.H.R., Goosen, N.K., Herman, P.M.J., Kromkamp, J., Middleburg, J.J., Soetaert, K., 1995. Production and consumption of biological particles in temperate tidal estuaries. Oceanography and Marine Biology: An Annual Review 33, 1–149.

Heron, S.F., Ridd, P.V., 2003. The effect of water density variations on the tidal flushing of animal burrows. Estuarine, Coastal and Shelf Science 58, 137–145.

Hilborn, R., Mangel, M., 1997. The Ecological Detective. Confronting Models with Data. Princeton University Press, Princeton, NJ, 315 pp.

Hinata, H., 2006. Effects of oceanic bottom water intrusion on the Tokyo Bay environment. pp. 67–78 in Wolanski, E. (ed.), The Environment in Asia Pacific Harbours. Springer, Dordrecht.

Hoa, L.T.V., Nhan, N.H., Wolanski, E., Cong, T.T., Shigeko, H., 2007. The combined impact on the flooding in the Mekong River delta, Vietnam, of local man-made structures, sea level rise, and dams upstream in the river catchment. Estuarine, Coastal and Shelf Science 71, 110–116.

Hodge, A.T., 2002. Roman Aqueducts and Water Supply. Duckworth Publishers, London. 504 pp.

Hoegh-Guldberg, O., 2004. Coral reefs in a century of rapid environmental change. Symbiosis 37, 1–31.

Holland, A., Zingmark, R., Dean, J., 1974. Quantitative evidence concerning the stabilization of sediments by marine benthic diatoms. Marine Biology 27, 191–196.

Hopkinson, C.S., Vallino, J.J., 2005. Efficient export of carbon to the deep ocean through dissolved organic matter. Nature 433, 142–145.

Horne, A.J., Javornicky, P., Goldman, C.R., 1971. A Freshwater 'Red Tide' on Clear Lake, California. Limnology and Oceanography 16, 684–689.

Hughes, T.P., Baird, A.H., Bellwood, D.R., Card, M., Connolly, S.R., Folke, C., Grosberg, R., Hoegh-Guldberg, O., Jackson, J.B.C., Kleypas, J., Lough, J.M., Marshall, P., Nyström, M., Palumbi, S.R., Pandolfi, J.M., Rosen, B., Roughgarden, J., 2003. Climate change, human impacts, and the resilience of coral reefs. Science 301, 929–933.

Huzzey, L.M., 1988. The lateral density distribution in a partially mixed estuary. Estuarine, Coastal and Shelf Science 27, 351–358.

Ikeda, S., Nakata, H., 2004. Systems analyses on the mechanism of 'red tide' outbreaks. pp. 395–432 in Okaichi, T. (ed.), Red Tides. Terra Scientific Publishing Company, Kluwer Academic Publishers, Tokyo.

Imboden, D.M., 1974. Phosphorus model of lake eutrophication. Limnology and Oceanography 19, 297–304.

Ippen, A.T., 1966. Estuary and Coastline Hydrodynamics. McGraw-Hill, New York, 744 pp.

Jacinto, G.S., Velasquez, I.B., San Diego-McGlone, M.L., Villanoy, C.L., Siringan, F.B., 2006. Biophysical environment of Manila Bay – then and now. pp. 293–307 in Wolanski, E. (ed.), The Environment in Asia Pacific Harbours, Springer, Dordrecht.

Jennings, S., Nicholson, M.D., Dinmore, T.A., Lancaster, J.E., 2002. Effects of chronic trawling disturbance on the production of infaunal communities. Marine Ecology Progress Series 243, 251–260.

Jiang, J., Wolanski, E., 1998. Vertical mixing by internal waves breaking at the lutocline, Jiaojiang River estuary, China. Journal of Coastal Research 14, 1426–1431.

Johnston, P.J., Lambeck, K., 1999. Postglacial rebound and sea level contributions to changes in the geoid and the earth's rotation axis. Geophysical Journal International 136, 537–558.

Jorgensen, S.E., Bendoricchio, G., 2001. Fundamentals of Ecological Modelling. Elsevier, Amsterdam, 530 pp.

Jorgensen, S.E., Kamp-Nielsen, L., Christensen, T., Windolf-Nielsen, J., Westergaard, B., 1986. Validation of a prognosis based upon an eutrophication model. Ecological Modelling 32, 165–182.

Ke, X., Collins, M.B., Poulos, S.E., 1994. Velocity structure and sea bed roughness associated with intertidal (sand and mud) flats and saltmarshes of the Wash, U.K. Journal of Coastal Research 10, 702–715.

Kerr, G., Nisbet, T.R., 1996. The Restoration of Floodplain Woodlands in Lowland Britain: a Scoping Study and Recommendations for Research. R&D Technical Report W15, Forestry Commission Research Division, 28 pp.

Keulegan, G.H., 1967. Tidal flow in entrances, water-level fluctuations of basins in communication with seas. Committee on Tidal Hydraulics, Corps of Engineers, U.S. Army, Vicksburg, MS.

Kim, T.I., Choi, B.H., Lee, S.W., 2006. Hydrodynamics and sedimentation induced by large-scale coastal developments in the Keum River estuary, Korea. Estuarine, Coastal and Shelf Science 68, 515–528.

King, B., Wolanski, E., 1996. Bottom friction reduction in turbid estuaries. American Geophysical Union. Mixing in Estuaries and Coastal Seas, Coastal and Estuarine Studies 50, 325–337.

Kingsford, M.J., 1990. Linear oceanographic features: a focus for research on recruitment processes. Australian Journal of Ecology 15, 391–401.

Kingsford, M.J., 1993. Biotic and abiotic structure in the pelagic environment: importance to small fish. Bulletin of Marine Science 53, 393–415.

Kirby, C.J., Gosselink, J.G., 1976. Primary production in a Louisiana Gulf Coast *Spartina alterniflora* marsh. Ecology 57, 1052–1059.

Kishi, M.J., Ikeda, S., 1986. Population dynamics of 'red tide' organisms in eutrophicated coastal water – numerical experiment of phytoplankton bloom in the East Seto Inland Sea, Japan. Ecological Modelling 31, 145–174.

Kitheka, J.U., 1997. Coastal tidally-driven circulation and the role of water exchange in the link between tropical coastal ecosystems. Estuarine, Coastal and Shelf Science 44, 177–187.

Kjerfve, B., 1986. Comparative oceanography of coastal lagoons. pp. 63–81 in Wolfe, D.A. (ed.), Estuarine Variability. Academic Press, New York.

Kjerfve, B., 1989. Estuarine geomorphology and physical oceanography. pp. 47–78 in Day, Jr W., Hall, C.A.S., Kemp, W.M., Yanez-Arancibia, A. (eds), Estuarine Ecology. Wiley, New York.

Kjerfve, B., Miranda, L.B., Wolanski, E., 1991. Modelling water circulation in an estuary and intertidal salt marsh system. Netherlands Journal of Sea Research 28, 141–147.

Kot, M., 2001. Elements of Mathematical Ecology. Cambridge University Press, Cambridge, 464 pp.

Laegdsgaard, P., 2006. Ecology, disturbance and restoration of coastal saltmarsh in Australia: a review. Wetlands Ecology and Management 14, 379–399.

Lambeck, K., Chappell, J., 2001. Sea level change through the last glacial cycle. Science 292, 679–686.

Lancelot, C., Staneva, J., Van Eeckhout, D., Beckers, J.M., Stanev, E., 2002. Modeling the response of the northwestern Black Sea ecosystem to changes in nutrient delivery by the Danube River after its damming in 1972. Estuarine, Coastal and Shelf Science 54, 473–499.

Lane, A., Prandle, D., 2006. Random-walk particle modelling for estimating bathymetric evolution of an estuary. Estuarine, Coastal and Shelf Science 68, 175–187.

Langlois, E., Bonis, A., Bouzille, J.B., 2001. The response of *Puccinellia maritima* to burial: a key to understanding its role in salt marsh dynamics? Journal of Vegetation Science 12, 289–297.

Langlois, E., Bonis, A., Bouzille, J.B., 2003. Sediment and plant dynamics in saltmarshes pioneer zone: *Puccinellia maritima* as a key species? Estuarine, Coastal and Shelf Science 56, 239–249.

Lara, R.J., Szlafsztein, C.F., Cohen, M.C.L., Berger, U., Glaser, M., 2002. Implications of mangrove dynamics for private land use in Bragança, North Brazil: a case study. Journal of Coastal Conservation 8, 97–102.

Largier, J., 2004. The importance of retention zones in the dispersal of larvae. American Fisheries Society Symposium 42, 105–122.

Leeuw, J. de, Dool, van den, Munck, W. de, Nieuwenhuize, J., Beeftink, W.G.,1991. Factors influencing the soil salinity regime along an intertidal gradient. Estuarine, Coastal and Shelf Science 32, 87–97.

Nienhuis, P.H., 1993. Nutrient cycling and foodwebs in Dutch estuaries. Hydrobiologia 265, 15–44.

Nobre, A.M., Ferreira, A.M., Newton, A., Simas, T., Icely, J.D., Neves, R., 2005. Management of coastal eutrophication: integration of field data, ecosystem-scale simulations and screening models. Journal of Marine Systems 56, 375–390.

Nunes, R.A., Simpson, J.H., 1985. Axial convergence in a well-mixed estuary. Estuarine, Coastal and Shelf Science 20, 637–649.

Nunes, R.A., Lennon, G.W., 1986. Physical property distributions and seasonal trends in Spencer Gulf, South Australia: an inverse estuary. Australian Journal of Marine and Freshwater Research 37, 39–53.

O'Brien, M.P., 1931. Estuary tidal prism related to entrance areas. Civil Engineering 1, 738–739.

Odum, W.E., Heald, E.J., 1972. Trophic analyses of an estuarine mangrove community. Bulletin of Marine Science 22, 671–738.

Okaichi, T., Yanagi, T., 1997. Sustainable Development in the Seto Inland Sea, Japan – From the Viewpoint of Fisheries. Terra Scientific Publishing Company. Tokyo, 329 pp.

Okubo, A., 1973. Effect of shoreline irregularities on streamwise dispersion in estuaries and other embayments. Netherlands Journal of Sea Research 6, 213–224.

Oldfield, F., Richardson, N., Appleby, P.G., Yu, L., 1993. ^{241}Am and ^{137}Cs activity in fine grained sediments from parts of the N.E. Irish Sea shoreline. Journal of Environmental Radioactivity 19, 1–24.

Onuf, C.P., 1994. Seagrasses, dredging and light in Laguna Madrea, Texas, U.S.A. Estuarine, Coastal and Shelf Science 39, 75–92.

Paerl, H.W., 1988. Nuisance phytoplankton blooms in coastal, estuarine, and inland waters. Part 2: comparative ecology of freshwater and marine ecosystems Limnology and Oceanography 33, 823–847.

Pages, J., Lemoalle, L., Fritz, B., 1995. Distribution of carbon in a tropical hypersaline estuary, the Casamance (Senegal, West Africa). Estuaries 18, 456–468.

Paling, E.I., van Keulen, M., Wheeler, K.D., 2001. Improving mechanical seagrass transplantation. Ecological Engineering 18, 107–113.

Parish, J.D., 1989. Fish communities of interacting shallow-water habitats in tropical oceanic regions. Marine Ecology Progress Series 58, 143–160.

Partheniades, E., 1965. Erosion and deposition of cohesive soils. Journal of the Hydraulics Division, American Society of Civil Engineers 91, 105–138.

Pendleton, E.A., FitzGerald, D.M., 2005. Comparison of the hydrodynamic character of three tidal inlet systems. pp. 81–100 in FitzGerald, D.M., Knight, J. (eds), High Resolution Morphodynamics and Sedimentary Evolution of Estuaries. Springer, Dordrecht.

Perillo, G.M.E., 1995. Definition and geomorphologic classifications of estuaries. pp. 17–47 in Perillo, G.M.E. (ed.), Geomorphology and Sedimentology of Estuaries. Developments in Sedimentology 53. Elsevier Science BV, Amsterdam.

Perillo, G., Minkoff, D.R., Piccolo, M.C., 2005. Novel mechanism of stream formation in coastal wetlands by crab–fish–groundwater interaction. Geo-Marine Letters 25, 214–220

Perillo, G.M.E., 1996. Geomorphology and Sedimentology of Estuaries. Elsevier, Amsterdam, 488 pp.

Perin, G., Fabris, R., Manente, S., Rebello Wagener, A., Hamacher, C., Scotto, S., 1997. A five-year study on the heavy metal pollution of Guanabra Bay sediments (Rio de Janeiro, Brazil), and evaluation of the metal bioavailability by means of geochemical speciation. Water Research 31, 3017–3028.

Peterken, G.F., Hughes, F.M.R., 1995. Restoration of floodplain forests in Britain. Forestry 68, 187–202.

Pethick, J.S., 1986. An introduction to coastal geomorphology. Edward Arnold, London, 260 pp.

Pethick, J.S., 1992. Saltmarsh geomorphology. pp 41–62 in Allen, J.R.L., Pye, K. (eds), Saltmarshes: Morphodynamics, Conservation and Engineering Significance. Cambridge University Press, Cambridge.

Pethick, J., Leggett, D., Husain, L., 1992. Boundary layers under salt marsh vegetation developed in tidal currents. pp. 113–124 in Thorne, J. (ed.), Vegetation and Erosion Processes and Environments. John Wiley, London.

Pirazzoli, P.A., 1991. World Atlas of Holocene Sea Level Changes. Elsevier, Amsterdam, 300 pp.

Postel, S., Richter, B., 2003. Rivers for Life: Managing Water for People and Nature. Island Press, Washington, DC, 220 pp.

Postma, H., 1967. Sediment transport and sedimentation in the estuarine environment. pp. 158–179 in Lauff, G.H. (ed.), Estuaries. American Association for the Advancement of Science, Publication No. 83.

Powell, E.N., Kraeuter, J.N., Ashton-Alcox, I.K.A., 2006. How long does oyster shell last on an oyster reef? Estuarine Coastal and Shelf Science 69, 531–542.

Prasetya, G.S., 2007. The role of coastal forest and trees in combating coastal erosion. pp. 100–127 in Braatz, S., Fortuna, S., Broadhead, J., Leslie, R. (eds). Coastal Protection in the Aftermath of the Indian Ocean Tsunami. What Role for Forests and Trees? FAO, Bangkok, 234 pp.

Preuss, J., 2007. Coastal area planning and management using forests and trees as protection from tsunamis. pp. 129–154 in Braatz, S., Fortuna, S., Broadhead, J., Leslie, R. (eds). Coastal Protection in the Aftermath of the Indian Ocean Tsunami. What Role for Forests and Trees? FAO, Bangkok, 234 pp.

Price, J.M., Reed, M., 2006. Merging engineering and science in marine environmental model applications. Estuarine, Coastal and Shelf Science 70, 523–524.

Pritchard, D.W., 1967. What is an estuary: a physical viewpoint. pp. 3–5 in Lauff, G.H. (ed.), Estuaries. Publication No. 83, American Association for the Advancement of Science, Washington, DC.

Rabalais, N.N., Turner, R.E., 2001. Hypoxia in the Northern Gulf of Mexico: description, causes and change. pp. 1–36 in Rabalais, N.N., Turner, R.E. (eds), Coastal Hypoxia: Consequences for Living Resources and Ecosystems. Coastal and Estuarine Studies 58, American Geophysical Union, Washington, DC.

Rabalais, N.N., Turner, E., Scavia, D., 2002. Beyond science into policy: Gulf of Mexico hypoxia and the Mississippi River. Bioscience 52, 129–142.

Radach, G., Moll, A., 2006. Review of three-dimensional ecological modeling related to the North Sea shelf system. Part II: model validation and data needs. Oceanography and Marine Biology: An Annual Review 44, 1–60.

Raudkivi, A.J., 1967. Loose boundary hydraulics. Pergamon Press, Oxford, 331 pp.

Ray, G.C., 1996. Coastal–marine discontinuities and synergisms: implications for biodiversity conservation. Biodiversity and Conservation 5, 1095–1108.

Reiter, P., 2000. From Shakespeare to Defoe: malaria in England in the Little Ice Age. Emerging Infectious Diseases 6, 1–11.

Rey, J.R., Kain, T., Crossman, R., Peterson, M., Shaffer, J., Vose, F., 1991. Zooplankton of impounded marshes and shallow areas of a subtropical lagoon. Florida Scientist 54, 191–203.

Ribbe, J., 2006. A study into the export of saline water from Hervey Bay, Australia. Estuarine, Coastal and Shelf Science 66, 550–558.

Richards, F.J., 1934. The salt marshes of the Dovey Estuary. IV. The rates of vertical accretion, horizontal extension and scarp erosion. Annals of Botany 48, 225–259.

Richardson, K., Jorgensen, B.B., 1996. Eutrophication: definition, history, and effects. pp 1–19, in Jorgensen, B.B., Richardson, K. (eds), Eutrophication in Coastal Marine Ecosystems. American Geophysical Union, Washington, DC.

Richmond, R.H., 2005. Recovering populations and restoring ecosystems: restoration of coral reefs and related marine communities. pp. 393–409 in Norse, E., Crowder, L. (eds), Marine Conservation Biology: The Science of Maintaining the Sea's Biodiversity. Island Press, Washington, DC.

Ridd, P.V., Sam, R., 1996. Profiling groundwater salt concentrations in mangrove swamps and tropical salt flats. Estuarine, Coastal and Shelf Science 43, 627–635.

Ridd, P., Sandstrom, M., Wolanski, E., 1988. Outwelling from tropical tidal salt flats. Estuarine, Coastal and Shelf Science 26, 243–253.

Riethmüller, R., Heineke, M., Küh, H., Keuker-Rüdiger, R., 2000. Chlorophyll a concentration as an index of sediment surface stabilization by microphytobenthos? Continental Shelf Research 20, 1351–1372.

Rijnsdorp, A.D., Buys, A.M., Storbeck, F., Visser, E.G., 1998. Micro-scale distribution of beam trawl effort in the southern North Sea between 1993 and 1996 in relation to the trawling frequency of the sea bed and the impact on benthic organisms. ICES Journal of Marine Science 55, 403–419.

Riley, R.W., Kent, C.P.S., 1999. Riley encased methodology: principles and processes of mangrove habitat creation and restoration. Mangroves and Salt Marshes 3, 207–213.

Robson, B.J., Hamilton, D.P., 2003. Summer flow event induces a cyanobacterial bloom in a seasonal Western Australia estuary. Marine and Freshwater Research 54, 139–151.

Roelke, D., Buyukates, Y., 2001. The diversity of harmful algal bloom-triggering mechanisms and the complexity of bloom initiation. Human and Ecological Risk Assessment 7, 1347–1362.

Roman, M.R., Holliday, D.V., Sanford, L.P., 2001. Temporal and spatial patterns of zooplankton in the Chesapeake Bay turbidity maximum. Marine Ecology Progress Series 213, 215–227.

Saenger, P., 2002. Mangrove Ecology, Silviculture and Conservation. Kluwer Academic Publishers, Dordrecht, 360 pp.

Salomons, W., Forstner, U., 1984. Metals in the Hydrocycle. Springer-Verlag, Berlin, 349 pp.

Sam, R., Ridd, P.V., 1998. Spatial variation of groundwater salinity in a mangrove-salt flat system, Cocoa Creek, Australia. Mangroves and Salt Marshes 2, 121–132.

Sanchez, J.M., San Leon, D.G., Izco, J., 2001. Primary colonisation of mudflats estuaries by *Spartina maritima* (Curtis) Fernald in Northwest Spain: vegetation structure and sediment accretion. Journal of Vegetation Science 7, 695–702.

Sanders, R.W., Wickham, S.A., 1993. Planktonic protozoa and metazoa: predation, food quality and population control. Marine Microbial Food Webs 7, 197–223.

Sanford, N.P., Dickhudt, P.J., Rubiano-Gomez, L., Yates, M., Suttles, S.E., Friedrichs, C.T., Fugate, D.D., Romine, H., 2005. Variability of suspended particle concentrations, sizes, and settling velocities in the Chesapeake Bay turbidity maximum. pp. 211–239 in Droppo, I.G., Leppard, G.G., Liss, S.N., Milligan, T.G. (eds), Flocculation in Natural and Engineered Environmental Systems. CRC Press, Boca Raton FL.

Savenije, H.H.G., 2005. Salinity and Tides in Alluvial Estuaries. Elsevier, Amsterdam, 194 pp.

Schoellhamer, D.H., 1996. Anthropogenic sediment resuspension mechanisms in a shallow microtidal estuary. Estuarine, Coastal and Shelf Science 43, 553–548.

Scholten, M., Rozema, J., 1990. The competitive ability of *Spartina anglica* on Dutch salt marshes. pp. 39–47 in Gray, A., Benham, P. (eds), *Spartina anglica* – A Research Review. Institute of Terrestrial Ecology, HMSO, London, 80 pp.

Schories, D., Barletta Bergan, A., Barletta, M., Krumme, U., Mehlig, U., Rademaker, V., 2003. The keystone role of leaf-removing crabs in mangrove forests of North Brazil. Wetlands Ecology and Management 11, 243–255.

Shabman, L., Scodari, P., 2004. Past, Present, and Future of Wetlands Credit Sales. Resources for The Future, Discussion Paper 04–48, Washington, DC, 36 pp.

Simpson, J.H., Turrell, W.R., 1986. Convergent fronts in the circulation of tidal estuaries. pp. 139–152 in Woolfe, D.A. (ed.), Estuarine Variability. Academic Press, London.

Sklar, F., Browder, J., 1998. Coastal environmental impacts brought about by alterations to freshwater flow in the Gulf of Mexico. Environmental Management 22, 547–562.

Smayda, T.J., 1997. Harmful algal blooms: their ecophysiology and general relevance to phytoplankton blooms in the sea. Limnology and Oceanography 42, 1137–1153.

Soukup, M.A., Portnoy, J.W., 1986. Impacts from mosquito control-induced sulfur mobilization in a Cape Cod Estuary. Environmental Conservation 13, 47–50.

Spalding, M.D., Ravilious, C., Green, E.P., 2001. World Atlas of Coral Reefs. University of California Press, Berkeley, CA, 424 pp.

Stommel, H., Farmer, H.G., 1952. On the Nature of Estuarine Circulation. WHOI Tech. Rep. 52–88, 131 pp.

Strand, M.N. 1997. Wetlands Deskbook. Environmental Law Institute, Washington, DC, 884 pp.

Streever, W.J., 2000. *Spartina alternifora* marshes on dredged material: a critical review of the ongoing debate over success. Wetlands Ecology and Management 8, 295–316.

Stumpf, R.P., Culver, M.E., Tester, P.A., Tomlinson, M., Kirkpatrick, G.J., Pederson, B.A., Truby, E., Ransibrahmanakul, V., Soracco, M., 2003. Monitoring *Karenia brevis* blooms in the Gulf of Mexico using satellite ocean color imagery and other data. Harmful Algae 2, 147–160.

Sutherland, T.F., Grant, J., Amos, C.L., 1998a. The effect of carbohydrate production by the diatom *Nitzshia curvilineata* on the erodibility of sediment. Limnology and Oceanography 43, 65–72.

Sutherland, T.F., Amos, C.L., Grant, J., 1998b. The effect of biofilms on the erodibility of sublittoral sediments of a temperate microtidal estuary. Limnology and Oceanography 43, 225–235.

Susilo, A., Ridd, P.V., 2005. The bulk hydraulic conductivity of mangrove soil perforated with animal burrows. Wetlands Ecology and Management 13, 123–133.

Susilo, A., Ridd, P.V., Thomas, S., 2005. Comparison between tidally driven groundwater flow and flushing of animal burrows in tropical mangrove swamps. Wetlands Ecology and Management 15, 377–388.

Syvitski, J.P.M., Harvey, N., Wolanski, E., Burnett, W.C., Perillo, G.M.E., Gornitz, V., 2005. Dynamics of the coastal zone. pp. 39–94 in Crossland, C.J., Kremer, H.H., Lindeboom, H.J., Marshall Crossland, J.I., Le Tissier, M.D.A. (eds), Coastal Fluxes in the Anthropocene. Springer-Verlag, Berlin, 231 pp.

Takahasi, H., Hosokawa, Y., Furukawa, K., Yoshimura, H., 2000. Visualisation of an oxygen-deficient bottom water circulation in Osaka Bay, Japan. Estuarine, Coastal and Shelf Science 50, 81–83.

Tenore, K.R., 1989. Some ecological perspectives in the study of nutrition of deposit feeders. pp. 309–317 in Lopez, G., Taghon, G., Levinton, J. (eds), Ecology of Marine Deposit Feeders. Springer-Verlag, New York.

Thompson, J.D., 1990. *Spartina anglica*, characteristic feature, or invasive weed of coastal salt marshes? Biologist 37, 9–12.

Thompson, R.C., Olsen, Y., Mitchell, R.P., Davis, A., Rowland, S.J., John, A.W.G., McGonigle, A., Russell, A.E., 2004. Lost at sea: where is all the plastic? Science 304, 838.

Toby, T., 1999. Nitrogen and phosphorus on oceanic primary production. Nature 400, 525–531.

Toggweiler, J.R., 1999. An ultimate limiting nutrient. Nature 400, 511–512.

Tomlinson, M.C., Stumpf, R.P., Ransibrahmanakul, V., Truby, E.W., Kirkpatrick, G.J., Pederson, B.A., Vargo, G.A., Heil, C.A., 2004. Evaluation of the use of SeaWiFS imagery for detecting *Karenia brevis* harmful algal blooms in the eastern Gulf of Mexico. Remote Sensing of Environment 91, 293–303.

Trimmer, M., Petersen, J., Mills, C., Sivyer, D., Young, E., Parker, R., 2005. Impact of benthic trawl disturbance on sediment biogeochemistry in the Southern North Sea. Marine Ecology Progress Series 298, 79–94.

Turner, R.E., 1993. Carbon, nitrogen and phosphorus leaching rates from *Spartina alterniflora* salt marshes. Marine Ecology Progress Series 92, 135–140.

Twilley, R.R., Lugo, A.E., Patterson-Zucca, C., 1986. Litter production and turnover in basin mangrove forests in southwest Florida. Ecology 67, 670–683.

Tyson, P., 1995. Sedimentary Organic Matter. Chapman & Hall, London, 615 pp.

Uijttewaal, W.S.J., Booij, R., 2000. Effects of shallowness on the development of free-surface mixing layers. Physics of Fluids 12, 392–402.

Uncles, R.J., Joint, I., Stephens, J.A. 1998a. Transport and retention of suspended particulate matter and bacteria in the Humber-Ouse estuary, United Kingdom, and their relationship to hypoxia and anoxia. Estuaries 21, 597–612.

Uncles, R.J., Howland, R.J.M., Easton, A.E., Griffiths, M.L., Harris, C., King, R.S., Morris, A.W., Plummer, D.H., Woodward, E.M.S., 1998b. Seasonal variability of dissolved nutrients in the Humber-Ouse Estuary, UK. Marine Pollution Bulletin 37, 234–246.

Uncles, R.J., Stephens, J.A., Smith, R.E., 2002. The dependence of estuarine turbidity on tidal intrusion length, tidal range and residence time. Continental Shelf Research 22, 1835–1856.

Uncles, R.J., Stephens, J.A., Law, D.J., 2006. Turbidity maximum in the macrotidal, highly turbid Humber Estuary, UK: flocs, fluid mud, stationary suspensions and tidal bores. Estuarine, Coastal and Shelf Science 67, 30–52.

UNEP, 1999. Global Environment Outlook 2000. United Nations Environment Programme. Available at: http://www.unep.org/unep/state.htm.

UNESCO, 2006. Integrated Watershed Management: Ecohydrology and Phytotechnology. UN, Venice, 246 pp.

Ungureanu, V.G., Stanica, A., 2000. Impact of human activities on the evolution of the Romanian Black Sea beaches. Lakes and Reservoirs: Research and Management 5, 111–115.

U.S. E.P.A. (U.S. Environmental Protection Agency), 2001. The U.S. Experience with Economic Incentives for Protecting the Environment. National Center for Environmental Economics, Report No. EPA 240-R-01-001, Washington, DC.

U.S. E.P.A. (U.S. Environmental Protection Agency (2002). The State of the Chesapeake Bay. A Report to the Citizens of the Bay Region. Report No. EPA 903-R-02-002, Annapolis, MD, 61 pp.

Valette-Silver, N.J., 1993. The use of sediment cores to reconstruct historical trends in contamination of estuarine and coastal sediments. Estuaries 16, 577–588.

Valiela, I. Cole, M.L., 2002. Comparative evidence that salt marshes and mangroves may protect seagrass meadows from land-derived Nitrogen loads. Ecosystems 5, 92–102.

Valle-Levinson, A., Moraga-Opazo, J., 2006. Observations of bipolar residual circulation in two equatorward-facing semiarid bays. Continental Shelf Research 26, 179–193.

van de Kreeke, J., 1988. Hydrodynamics of tidal inlets. pp. 1–23 in Aubrey, D.G., Weishar, L. (eds), Hydrodynamics and sediment dynamics of tidal inlets. Lecture Notes on Coastal and Estuarine Studies, Vol. 29. Springer-Verlag, New York.

van der Valk, A.G., 2006. The Biology of freshwater wetlands. Oxford University Press, Oxford, 173 pp.

Victor, S., Golbuu, Y., Wolanski, E., Richmond, R., 2004. Fine sediment trapping in two mangrove-fringed estuaries exposed to contrasting land-use intensity, Palau, Micronesia. Wetlands Ecology and Management 12, 277–283.

Victor, S., Neth, L., Golbuu, Y., Wolanski, E., Richmond, R.H., 2006. Sedimentation in mangroves and coral reefs in a wet tropical island, Pohnpei, Micronesia. Estuarine Coastal and Shelf Science 66, 409–416.

Walbridge, M.R., Struthers, J.P., 1993. Phosphorus retention in non-tidal palustrine forested wetlands of the Mid-Atlantic Region. Wetlands 13, 84–94.

Wang, J., Seliskar, D.M., Gallagher, J.L., League, M.T., 2006. Blocking *Phragmites australis* reinvasion of restored marshes using plants selected from wild populations and tissue culture. Wetlands Ecology and Management 14, 539–547.

Wattayakorn, G., Wolanski, E., Kjerfve, B., 1990. Mixing, trapping and outwelling in the Klong Ngao mangrove swamp, Thailand. Estuarine, Coastal and Shelf Science 31, 667–688.

Wei, T., Chen, Z., Duan, L., Gu, J., Saito, Y., Zhang, W., Wang, Y., Kanai, Y., 2007. Sedimentation rates in relation to sedimentary processes of the Yangtze Estuary, China. Estuarine, Coastal and Shelf Science 71, 37–46.

Wells, M.G., van Heijst, G.-J.F., 2003. A model of tidal flushing of an estuary by dipole formation. Dynamics of Atmospheres and Oceans 37, 223–244.

West, J.M., Zedler, J.B., 2000. Marsh–creek connectivity: fish use of a tidal salt marsh in southern California. Estuaries 23, 699–710.

WHO, 2000. Climate change and health: impact and adaptation. Available at http://www.who.int/environmental_information/Climate/climchange.pdf.

Williams, T.B., Bubb, J.M., Lester, J.N., 1994. Metal accumulation within salt marsh environments: a review. Marine Pollution Bulletin 28, 277–290.

Williams, D., Wolanski, E., Spagnol, S., 2006. Hydrodynamics of Darwin Harbour. pp. 461–476 in Wolanski, E. (ed.), The Environment in Asia Pacific Harbours. Springer, Dordrecht.

Winterwerp, J.C., Van Kesteren, W.G.M., 2004. Introduction to the physics of cohesive sediments in the marine environment. Developments in Sedimentology 56, Elsevier, Amsterdam, 466 pp.

Wolanski, E., 1986. An evaporation-driven salinity maximum zone in Australian tropical estuaries. Estuarine, Coastal and Shelf Science 22, 415–424.

Wolanski, E., 1992. Hydrodynamics of mangrove swamps and their coastal water. Hydrobiologia 247, 141–146.

Wolanski, E., 1994. Physical Oceanographic Processes of the Great Barrier Reef. CRC Press, Boca Raton, FL, 194 pp.

Wolanski, E., 2001. Oceanographic Processes of Coral Reefs. Physical and Biological Links in the Great Barrier Reef. CRC Press, Boca Raton, FL, 356 pp.

Wolanski, E., 2006a. The Environment in Asia Pacific Harbours. Springer, Dordrecht, 497 pp.

Wolanski, E., 2006b. The evolution time scale of macro-tidal estuaries: examples from the Pacific Rim. Estuarine, Coastal and Shelf Science 66, 544–549.

Wolanski, E., 2007. Protective functions of coastal forests and trees against natural hazards. pp. 157–179 in Braatz, S., Fortuna, S., Broadhead, J., Leslie, R. (eds), Coastal Protection in the Aftermath of the Indian Ocean Tsunami. What Role for Forests and Trees? FAO, Bangkok, 234 pp.

Wolanski, E., De'ath, G., 2005. Predicting the present and future human impact on the Great Barrier Reef. Estuarine, Coastal and Shelf Science 64, 504–508.

Wolanski, E., Duke, N., 2002. Mud threat to the Great Barrier Reef of Australia. pp. 533–542 in Healy, T., Wang, Y., Healy, J.A. (eds), Muddy Coasts of the World: Processes, Deposits and Function. Elsevier Science B.V., Amsterdam.

Wolanski, E., Gardiner, R., 1981. Flushing of salt from mangroves. Australian Journal Marine Freshwater Research 32, 681–683.

Wolanski, E., Hamner, W., 1988. Topographically controlled fronts and their biological influence. Science 241, 177–181.

Wolanski, E., Imberger, J., 1987. Friction-controlled selective withdrawal near inlets. Estuarine, Coastal and Shelf Science 24, 327–333.

Wolanski, E., Ridd, P., 1986. Tidal mixing and trapping in mangrove swamps. Estuarine, Coastal Shelf Science 23, 759–771.

Wolanski, E., Ridd, P., 1990. Coastal trapping and mixing in tropical Australia. pp. 165–183 in Cheng, R.T. (ed.), Long-term currents and residual circulation in estuaries and coastal seas. Springer-Verlag, New York.

Wolanski, E., Spagnol, S., 2000. Environmental degradation by mud in tropical estuaries. Regional Environmental Change 1, 152–162.

Wolanski, E., Spagnol, S., 2003. Dynamics of the turbidity maximum in King Sound, tropical Western Australia. Estuarine, Coastal and Shelf Science 56, 877–890.

Wolanski, E., Jones, M., Bunt, J.S., 1980. Hydrodynamics of a tidal creek–mangrove swamp system. Australian Journal Marine Freshwater Research 31, 431–450.

Wolanski, E., Jones, M., Williams, W.T., 1981. Physical properties of Great Barrier Reef lagoon waters near Townsville. II. Seasonal variations. Australian Journal Marine Freshwater Research 32, 321–334.

Wolanski, E., Imberger, J., Heron, M., 1984. Island wakes in shallow coastal waters. Journal Geophysical Research 89(C6), 10553–10569.

Wolanski, E., Drew, E., Abel, K., O'Brien, J., 1988. Tidal jets, nutrient upwelling and their influence on the productivity of the algal *Halimeda* in the Ribbon Reefs, Great Barrier Reef. Estuarine, Coastal and Shelf Science 26, 169–201.

Wolanski, E., Gibbs, R., Mazda, Y., Mehta, A., King, B., 1992a. The role of buoyancy and turbulence in the settling of mud flocs. Journal of Coastal Research 8, 35–46.

Wolanski, E., Mazda, Y., Ridd, P.V., 1992b. Mangrove hydrodynamics. pp. 43–62 in Robertson, A.I., Alongi, D.M. (eds), Tropical Mangrove Ecosystems. Coastal and Estuarine Studies 41, American Geophysical Union, Washington, DC.

Wolanski, E., Asaeda, T., Tanaka, A., Deleersnijder, E., 1996a. Three-dimensional island wakes in the field, laboratory and numerical models. Continental Shelf Research 16, 1437–1452.

Wolanski, E., Huan, N., Dao, L., Nhan, L., Thuy, N., 1996b. Fine sediment dynamics in the Mekong River estuary, Vietnam. Estuarine, Coastal and Shelf Science 43, 565–582.

Wolanski, E., Spagnol, S., Ayukai, T., 1998. Field and model studies of the fate of particulate carbon in mangrove-fringed Hinchinbrook Channel, Australia. Mangroves and Salt Marshes 2, 205–221.

Wolanski, E., Spagnol, S., King, B., 1999. Patchiness in the Fly River plume, Papua New Guinea. Journal of Marine Systems 18, 369–381.

Wolanski, E., Mazda, Y., Furukawa, K., Ridd, P., Kitheka, J., Spagnol, S., Stieglitz, T., 2001. Water circulation through mangroves and its implications for biodiversity. pp. 53–76 in Wolanski, E. (ed.), Oceanographic Processes of Coral Reefs: Physical and Biological Links in the Great Barrier Reef. CRC Press, Boca Raton, Florida.

Wolanski, E., Richmond, R., McCook, L., Sweatman, H., 2003a. Mud, marine snow and coral reefs. American Scientist 91, 44–51.

Wolanski, E., Richmond, R.H., Davis, G., Bonito, V., 2003b. Water and fine sediment dynamics in transient river plumes in a small, reef-fringed bay, Guam. Estuarine, Coastal and Shelf Science 56, 1029–1043.

Wolanski, E., Boorman, L.A., Chicharo, L., Langlois-Saliou, E., Lara, R., Plater, A.J., Uncles, R.J., Zalewski, M., 2004a. Ecohydrology as a new tool for sustainable management of estuaries and coastal waters. Wetlands Ecology and Management 12, 235–276.

Wolanski, E., Richmond, R.H., McCook, L., 2004b. A model of the effects of land-based human activities on the health of coral reefs in the Great Barrier Reef and in Fouha Bay, Guam, Micronesia. Journal of Marine Systems, 46, 133–144.

Wolanski, E., Fabricius, K., Spagnol, S., Brinkman, R., 2005. Fine sediment budget on an inner-shelf coral-fringed island, Great Barrier Reef of Australia. Estuarine, Coastal and Shelf Science 65, 153–158.

Wolanski, E., Williams, D., Hanert, E., 2006a. The sediment trapping efficiency of the macro-tidal Daly Estuary, tropical Australia. Estuarine, Coastal and Shelf Science 69, 291–298.

Wolanski, E., Chicharo, L., Chicharo, M., Morais, P., 2006b. An ecohydrology model of the Guadiana Estuary (South Portugal). Estuarine, Coastal and Shelf Science 70, 132–143.

Wolanski, E., McKinnon, D., Alongi, D., Spagnol, S., Williams, D., 2006c. An ecohydrology model of Darwin Harbour. pp. 477–488 in Wolanski, E. (ed.), The Environment in Asia Pacific Harbours. Springer, Dordrecht, 497 pp.

Woodroffe, C.D., 2003. Coasts, Form, Processes and Evolution. Cambridge University Press, Cambridge, 623 pp.

Woodroffe, C.D., Chappell, J., Thom, B., Wallensky, E., 1986. Geomorphological Dynamics and Evolution of the South Alligator Tidal River and Plains, Northern Territory. North Australian Research Unit, Mangrove Monograph No. 3, Darwin, 190 pp.

Yamamoto, T., Seike, T., Hashimoto, T., Tarutani, K., 2002. Modelling the population dynamics of the toxic dinoflagellate *Alexandrium Tamarense* in Hiroshima Bay, Japan Journal of Plankton Research 24, 33–47.

Yanagi, T., Yamamoto, T., Koizumi, Y., Ikeda, T., Kamizono, M., Tamori, H., 1995. A numerical simulation of red tide formation. Journal of Marine Systems 6, 269–285.

Zaitsev, Y.P., 1992. Recent changes in the trophic structure of the Black Sea. Fisheries Oceanography 1, 180–189.

Zalewski, M., 2002. Ecohydrology – the use of ecological and hydrological processes for sustainable management of water resources. Hydrological Sciences Bulletin 47, 823–832.

Zhang, J., Yang, S.L., Xu, Z.L., Wu, Y., 2006. Impact of human activities on the health of ecosystems in the Changjiang delta region. pp. 93–111 in Wolanki, E. (ed.), The Environment in Asia Pacific Harbours. Springer, Dordrecht.